New Caribbean Molluscan Faunas

Library of Congress Catalog Card Number: QL408.2.P47
ISBN: 0-938415-01-8

92 91 90 89 88 87   5 4 3 2 1
Manufactured in the United States of America.

Library of Congress Cataloging in Publication Data
Petuch, Edward J.
New Caribbean molluscan faunas.

"A CERF Book."
Bibliography: p.
Includes index.
1. Mollusks—Caribbean Sea. I. Coastal Education and Research Foundation (U.S.) II. Title.
QL408.2.P47   1987   594.092'35   87-13842
ISBN 0-938415-01-8 (pbk.)

New Caribbean Molluscan Faunas

Edward J. Petuch
Florida Atlantic University
Boca Raton, Florida

The Coastal Education & Research Foundation ***[CERF]***
Charlottesville, Virginia

A **CERF** Book

DEDICATION

To my wife Linda, and my sons, Eric and Brian—who patiently tolerated, over so many months, my disappearance into the complex world of molluscan systematics.

TABLE OF CONTENTS

ILLUSTRATIONS

CHAPTER 1

New Age of Discovery

Throughout the late 18th and the 19th century, the Caribbean region yielded a continuous stream of new species of marine mollusks. These malacological discoveries caused great excitement in the European scientific community. Unfortunately, collecting techniques were primitive at the time, with most deep water species being collected in fish traps or nets. Still, the French West Indian colonies produced such rarities as *Pleurotomaria (Entemnotrochus) adansonianus* Crosse and Bernardi, 1861, *Cypraea aubryana* Jousseaume, 1869 (discussed later in this book), *Phalium coronadoi* (Crosse, 1867), *Lyria beaui* (Fischer and Bernardi, 1857), and *Conus mazei* Deshayes, 1874. The English colonies produced an equal number of major discoveries such as *Morum dennisoni* (Reeve, 1843), and *Chicoreus spectrum* (Reeve, 1846). This initial age of discovery showed that the geographically-small Caribbean Sea contained a molluscan fauna that rivaled that of the immense Indo-Pacific tropical region.

The end of the 19th century saw the single greatest spurt of new species ever assembled at one time for the Caribbean region. This amazing contribution to systematic malacology, which encompassed the description of literally hundreds of new mollusks, was the work of a single individual, William Healey Dall. Working with material collected through dredging by the steamer *Blake,* and other fisheries research vessels, Dall methodically documented the previously-unknown deep water faunas of the Gulf of Mexico, Florida coasts, and the Caribbean islands. Amongst his most important discoveries were the first known Caribbean abalone, *Haliotis pourtalesi* Dall, 1881, the first known Caribbean and Atlantic species of the Californian rock scallops, *Hinnites adamsi* Dall, 1886, and the first known Atlantic members of normally Japanese groups such as *Lischkeia (Bathybembix) imperialis* (Dall, 1881) and *Phenacovolva (Calcaria) piragua* (Dall, 1889). Also an authority on fossil mollusks, Dall named hundreds of new species from the Tertiary formations of Florida, the southeastern U. S., and the western coast of North America. This unbelievably prolific individual was, without doubt, the greatest American malacologist.

Following in the footsteps of Dall, a second wave of new Caribbean species descriptions appeared in the 1940's, 1950's, and early 1960's in the *Johnsonia* monograph series, under the tutelage of William Clench of Harvard University. Clench, himself, described many new and exiting species, including the first American-Caribbean species of Japonic genera, such as the volute *Teremachia mirabilis* (Clench and Aguayo), and the latiaxis *Tarantellaxis juliae* (Clench and Aguayo). Interestingly enough, with all this work and that of Dall, there were still four areas of the Caribbean Sea-Gulf of Mexico area that were only casually studied and were faunistically unknown. These were that continental shelf off western Florida (depths of 150-200 m), the Bahamas Platform, the Gulf of Honduras, and the Colombian and Venezuelan coasts of northern South America. Although the last Caribbean areas to be studied in detail, these four areas are now proving to contain some of the richest molluscan faunas in the western Atlantic, and faunas that are replete with new and undescribed species.

The new age of discovery in the Caribbean was issued in by the pioneer work of Frederick M. Bayer, then of the University of Miami (BAYER, 1971). During the 1960's, the University of Miami sent a series of marine collecting and faunal survey expeditions to the waters around the Bahamas, Gulf of Honduras, and northern coast of South America. Utilizing the research vessels *R/V John Elliot Pillsbury* and the *R/V Gerda,* the university undertook the first deep water trawling off these areas and brought to light an entirely new world of malacology. Bayer and his associates worked up this material and described some spectacular new animals, some new to science and some previously known from single or fragmentary specimens. Among some of the more important finds were an entire new fauna of *Columbarium (Fulgurofusus)* species, a new fauna of volutes, including a new endemic Caribbean volute genus, *Cordilyria (C. cordis* Bayer), and the Japonic-looking coralliophilids, *Babelomurex fax* Bayer and *Babelomurex sentix* Bayer.

The advent of SCUBA diving after World War II has also aided in the discovery of new Caribbean mollusks. SCUBA divers are able to collect on areas of reefs and offshore banks that were previously difficult or impossible to visit. They can also spend more time picking for small, cryptic species. Easy access to air travel into previously inaccessible areas and islands has allowed inspired amateur malacologists/divers to collect untouched molluscan faunas. Over the past decade, there has been an increasing number of new species of mollusks that have been accumulated by these amateurs from all over the Caribbean. These collectors are the vanguard of the new age of discovery in the Caribbean region, but sadly, there are few professional malacological systematists (specializing in molluscan taxonomy) to take advantage of these new and exciting discoveries. This is primarily due to the fact that most universities and marine laboratories no longer teach in-depth courses in taxonomy but, instead, emphasize more modern theoretical subjects. Because of this backlog of unnamed important marine animals, I have decided to take up some of the slack and describe a number of the more obvious new species in this book. In the tradition of Dall, Clench, and Bayer, it is hoped that my contribution will shed a little more light onto the faunal composition of one of the world's richest biotic regions.

Interestingly enough, there is a feeling amongst the current systematic malacological community in the United States that virtually every marine mollusk in the Western Atlantic has already been discovered and named. Indeed, as a young, fledgling malacologist who had just received his M.S. degree, I heard several speakers at the 1975 American Malacological Union meeting state repeatedly that the Age of Discovery for western Atlantic systematic malacology had ended with the death of William Healey Dall (*i.e.,* ROSEWATER, 1976). One well-known popularizer of conchology even went so far as to state that every mollusk in the Caribbean had been named at least twice. With these thoughts still fresh in mind, you can imagine my surprise at discovering entire new faunas within the Caribbean region when I began my doctoral research trips in 1976. It pleases me greatly to find that the arrogant pronouncements on knowing everything that there is to know about the systematics of Caribbean mollusks are totally false. I only hope that this little book will convey the excitement of a whole New Age of Discovery that we are now entering. Armed with this enthusiasm for discovery while utilizing new collecting techniques, a whole new generation of young scientists and inspired amateurs will be able to restore to Caribbean malacology the vitality that has suffered a hiatus over the last decade.

The chapters of this book are divided by biogeographical regions with the western Atlantic and their accompanying endemic faunas. The relatively unexplored areas covered in this book include: *1. Florida*—the deep water (200-300 m depth) areas along the edge of the continental shelf of western Florida in the Gulf of Mexico; the deep water areas off eastern Florida; the *Thalassia* (Turtle Grass) community within the Everglades National Park section of Florida Bay, and rocky intertidal regions along southeastern Florida; *2. The Bahamas*—the deep water areas (200-300 m depth) off

Grand Bahama Island; the shallow reef environment of the Bimini Chain; the reef environment of the Cay Sal Bank; *3. Eastern Honduras and Eastern Central America*—the offshore and reef environments of the Bay Islands (Roatan, Utila, and Guanaja); the reef environment of Cozumel Island, Mexico; the soft bottom habitats off southern Belize; the banks offshore of Belize, Honduras, Nicaragua, and Panama; *4. Northern South America*—the offshore (35-200 m depth) areas of the Gulf of Venezuela and Gulf of Triste, Venezuela; the reef-associated environments of Aruba and Bonaire Islands, Netherlands Antilles; the deep water (200-300 m depth) areas off Barbados; *5. Northern Brazil*—the offshore (10-35 m depth) area of Bahia; the reef environment of the Abrolhos Archipelago.

These areas, combined, have yielded hundreds of rare and seldom-seen mollusks and 107 new species, which are described here for the first time. These include:

NEW SPECIES FROM THE CARIBBEAN REGION with USNM (Smithsonian Institution) type depository numbers

Turbinidae
1. *Bolma (Senobolma) sunderlandi* n.sp., holotype USNM 859814
2. *Homalopoma finkli* n.sp., holotype USNM 859815

Capulidae
3. *Cyclotheca pacei* n.sp., holotype USNM 859816, paratype USNM 859817

Cerithiidae
4. *Cerithium eburneum aliceae* n.subsp., holotype USNM 859818, paratype USNM 859819
5. *Cerithium lindae* n.sp., holotype USNM 859820, paratypes USNM 859821
6. *Cerithium muscarum pacei* n.subsp., holotype USNM 859822, paratype USNM 859823

Modulidae
7. *Modulus kaicherae* n.sp., holotype USNM 859824
8. *Modulus lindae* n.sp., holotype USNM 859825
9. *Modulus pacei* n.sp., holotype USNM 859826, paratype USNM 859827

Turritellidae
10. *Turritella (Torcula) lindae* n.sp., holotype USNM 859802

Architectonicidae
11. *Architectonica sunderlandi* n.sp., holotype USNM 859906

Ovulidae
12. *Cyphoma lindae* n.sp., holotype USNM 859806
13. *Phenacovolva (Turbovula) lindae* n.sp., holotype USNM 859828
14. *Pseudocyphoma gibsonsmithorum* n.sp., holotype USNM 859829

Triviidae
15. *Trivia (Niveria) pacei* n.sp., holotype USNM 859830

Cypraeidae
16. *Cypraea (Siphocypraea) mus tristensis* n.subsp, holotype USNM 784455

Cassidae
17. *Morum (Cancellomorum) lindae* n.sp.,holotype USNM 859831
18. *Sconsia lindae* n.sp., holotype USNM 784457

Cymatiidae
19. *Distorsio robinsoni* n.sp., holotype USNM 859832, paratype USNM 859833

Muricidae
20. *Chicoreus dunni* n.sp., holotype USNM 859835
21. *Chicoreus emilyae* n.sp., holotype USNM 859834
22. *Chicoreus rachelcarsonae* n.sp., holotype USNM 859811
23. *Dermomurex (Trialatella) glicksteini* n.sp., holotype USNM 859836
24. *Dermomurex (Trialatella) kaicherae* n.sp., holotype USNM 859837
25. *Favartia lindae* n.sp., holotype USNM 859805
26. *Lindapterys sanderi* n.sp., holotype USNM 859838
27. *Lindapterys vokesae* n.sp., holotype (Paleobiology) USNM 647012
28. *Murex garciai* n.sp., holotype USNM 859839
29. *Murex samui* n.sp., holotype USNM 859840
30. *Murex sunderlandi* n.sp., USNM 859841
31. *Murexiella hilli* n.sp., holotype USNM 859842
32. *Murexiella kalafuti* n.sp., holotype USNM 859810
33. *Murexiella leonardhilli* n.sp., holotype USNM 859843
34. *Murexiella taylorae* n.sp., holotype USNM 859844
35. *Murexsul harasewychi* n.sp., holotype USNM 859845
36. *Muricopsis aliceae* n.sp., holotype USNM 859846
37. *Muricopsis richardbinghami* n.sp., holotype USNM 859847
38. *Muricopsis sunderlandi* n.sp., holotype USNM 859848, paratype USNM 859849

Coralliophilidae
39. *Coralliophila pacei* n.sp., holotype USNM 859850
40. *Quoyula kalafuti* n.sp., holotype USNM 859851

Fasciolariidae
41.*Harasewychia harasewychi* n.sp., holotype USNM 859852
42. *Heilprinia lindae* n.sp., holotype USNM 859808

Melongenidae-Busyconinae
43. *Busycon (Busycoarctum) lyonsi* n.sp., holotype USNM 859803, paratype USNM 859804

Colubrariidae
44. *Colubraria bayeri* n.sp., holotype USNM 859853

Buccinidae
45. *Antillophos bayeri* n.sp., holotype USNM 859854
46. *Mohnia kaicherae* n.sp., holotype USNM 859855

47. *Muricantharus harasewychi* n.sp., holotype USNM 859801
48. *Truncaria lindae* n.sp., holotype USNM 859856, paratype USNM 859857

Columbellidae
49. *Zafrona sunderlandi* n.sp., holotype USNM 859905
50. *Zafrona taylorae* n.sp., holotype USNM 859858, paratype USNM 859859

Olividae
51. *Agaronia hilli* n.sp., holotype USNM 859860
52. *Agaronia leonardhilli* n.sp., holotype USNM 859861
53. *Ancilla (Amalda) pacei* n.sp., holotype USNM 859862
54. *Gracilancilla lindae* n.sp., holotype USNM 859863
55. *Oliva bifasciata sunderlandi* n.subsp.,holotype USNM 859904
56. *Oliva circinata tostesi* n.subsp., holotype USNM 859865
57. *Oliva sargenti* n.sp., holotype USNM 859864
58. *Turrancilla williamsoni* n.sp., holotype USNM 859866

Mitridae
59. *Dibaphimitra janetae* n.sp., holotype USNM 859867
60. *Subcancilla leonardhilli* n.sp., holotype USNM 784574
61. *Subcancilla lindae* n.sp., holotype USNM 784575

Costellariidae
62. *Nodicostellaria kremerae* n.sp., holotype USNM 859868
63. *Turricostellaria leonardhilli* n.sp., holotype USNM 859869
64. *Turricostellaria lindae* n.sp., holotype USNM 859870

Volutomitridae
65. *Conomitra leonardhilli* n.sp., holotype USNM 784576
66. *Conomitra lindae* n.sp., holotype USNM 784577

Volutidae
67. *Falsilyria harasewychi* n.sp., holotype USNM 784486
68. *Falsilyria hilli* n.sp., holotype USNM 859871
69. *Falsilyria sunderlandi* n.sp., holotype USNM 859903
70. *Voluta lindae* n.sp., holotype USNM 859813

Cancellariidae
71. *Aphera lindae* n.sp., holotype USNM 784467
72. *Cancellaria richardpetiti* n.sp., holotype USNM 859809

Conidae
73. *Conus abrolhosensis* n.sp., holotype USNM 859872
74. *Conus kirkandersi* n.sp., holotype USNM 859873
75. *Conus aureonimbosus* n.sp., holotype USNM 859812
76. *Conus aureopunctatus* n.sp., holotype USNM 859874
77. *Conus bayeri* n.sp., holotype USNM 859875
78. *Conus binghamae* n.sp., holotype USNM 859876

79. *Conus colombianus* n.sp., holotype USNM 859877
80. *Conus eversoni* n.sp., holotype USNM 859878
81. *Conus finkli* n.sp., holotype USNM 859879
82. *Conus floridanus patglicksteinae* n.subsp., holotype USNM 859880
83. *Conus glicksteini* n.sp., holotype USNM 859881
84. *Conus harasewychi* n.sp., holotype USNM 859882
85. *Conus harlandi* n.sp., holotype USNM 859883
86. *Conus kevani* n.sp., holotype USNM 859884
87. *Conus leekremeri* n.sp., holotype USNM 859885
88. *Conus lindae* n.sp., holotype USNM 859886
89. *Conus magnottei* n.sp., holotype USNM 859887
90. *Conus pacei* n.sp., holotype USNM 859888
91. *Conus paraguana* n.sp., holotype USNM 859889
92. *Conus parascalaris* n.sp., holotype USNM 859890
93. *Conus perprotractus* n.sp., holotype USNM 859891
94. *Conus sunderlandi* n.sp., holotype USNM 859892
95. *Conus tristensis* n.sp., holotype USNM 784470
96. *Conus venezuelanus* n.sp., holotype USNM 784469

Terebridae
97. *Terebra (Myurella) lindae* n.sp., holotype USNM 859807
98. *Terebra (Strioterebrum) biminiensis* n.sp., holotype USNM 859893
99. *Terebra (Strioterebrum) pacei* n.sp., holotype USNM 859894

Turridae
100. *Hindsiclava tippetti* n.sp., holotype USNM 784649
101. *Paraborsonia lindae* n.sp., holotype USNM 784581
102. *Polystira lindae* n.sp., holotype USNM 859895
103. *Polystira sunderlandi* n.sp., holotype USNM 859896
104. *Splendrillia sunderlandi* n.sp., holotype USNM 859800

Haminoediae
105. *Haminoea taylorae* n.sp., holotype USNM 859897, paratypes USNM 859898

Bivalvia
Nuculanidae
106. *Scissuladrana ludmillae* n.sp., holotype USNM 859899, paratype USNM 859900

Pectinidae
107. *Argopecten irradians taylorae* n.subsp., holotype USNM 859901, paratypes USNM 859902

Five new genera are also described here for the first time:

New Genera From The Caribbean Region

Gastropoda:
Lindapterys n.gen., Family Muricidae. Type species: *Lindapterys vokesae* Petuch n.sp.

Harasewychia n.gen., Family Fasciolariidae. Type species: *Harasewychia harasewychi* Petuch n.sp.

Nodicostellaria n.gen., Family Costellariidae. Type species: *Costellaria kaicherae* Petuch, 1979.

Turricostellaria n.gen., Family Costellariidae. Type species: *Turricostellaria lindae* Petuch n.sp.

Bivalvia
Scissuladrana n.gen., Family Nuculanidae. Type species: *Scissuladrana ludmillae* Petuch n.sp.

Abbreviations used to denote the depository of type specimens of the new species are as follows:

USNM—Collection of the Division of Mollusks, National Museum of Natural History, Smithsonian Institution, Washington, DC.

DMNH—Collection of the Department of Malacology, Delaware Museum of Natural History, Greenville, Delaware.

RSMAS—Invertebrate collection, Rosenstiel School of Marine and Atmospheric Science, University of Miami, Miami, Florida.

ANSP—Collection of the Department of Malacology, Academy of Natural Sciences of Philadelphia, Philadelphia, Pennsylvania.

ACKNOWLEDGMENTS

I thank the following for the excellent photography used in this book: Mr. Kevan Sunderland, Ft. Lauderdale, Florida, Mrs. Sally Diana Kaicher, St. Petersburg, Florida, Dr. M. G. Harasewych, Department of Invertebrate Zoology, Smithsonian Institution, and Dr. Frederick M. Bayer, Department of Invertebrate Zoology, Smithsonian Institution. Without the generous help of these people, this book would never have been completed. Special thanks are given to the Rosenstiel School of Marine and Atmospheric Science, University of Miami, for loan of valuable research material and for allowing me the honor of being able to name some of the new species that were housed in their extensive and invaluable research collections. I also thank the Everglades National Park, Department of the Interior, for allowing me access to molluscan collections from Florida Bay. For the generous loans of research material and specimens for types, I thank Mr. Kevan Sunderland; Mr. Robert Pace, Miami, Florida; Mr. Wayne Harland, Lighthouse Point, Florida; Dr. and Mrs. Marvin Glickstein, Palm Beach Gardens, Florida; Mr. and Mrs. Richard Bingham, West Palm Beach, Florida; Continental Shelf, Inc., Tequesta, Florida (under contract with Shell Oil Corporation); Mr. Leonard Hill, Miami, Florida; Dr. Emilio Garcia, Lafayette, Louisiana; Florida Department of Natural Resources, St. Petersburg Marine Laboratory; Mr. Theodore Kalafut, Key West, Florida; Mr. Gene Everson, Ft. Lauderdale, Florida; Mr. David Robinson, Tulane University, New Orleans, Louisiana; Mr. Gary Magnotte, Pompano Beach, Florida; the late Mr. Kirk Anders, Ft. Lauderdale, Florida; the late Mr. Steve Samu, Palm Beach, Florida; Dr. and Mrs. Jack Gibson-Smith, Caracas, Venezuela; and the late Dr. Axel Olsson, Coral Gables, Florida. Very special thanks also go to Lee, Janet, and Raymond Kremer, Crystal Lake, Illinois.

Note: Many of my new taxa are described as patronyms. This was done in order to honor the people who have made so many important contributions to our knowledge of Caribbean molluscan fauna. I have also chosen to honor those who have given me support, both emotional and physical, when I needed it most during the course of my research.

CHAPTER 2

Geographical Heterochrony in the Caribbean

Instead of being a single large, homogenous faunal region, the Caribbean Sea is actually a collection of separate, geographically-discrete faunas packed into one area. Besides ubiquitous species such as *Strombus gigas, Strombus costatus, Cypraea zebra, Cassis tuberosa,* and *Conus spurius,* a great part of the Caribbean molluscan fauna is limited to disjunct areas around the Caribbean Basin. For example, if I were collecting cones, olives and volutes from intertidal sand flat localities at Bimini (Bahamas), Aruba (Dutch West Indies), and Cartagena (Colombia), I would find *Conus jaspideus, Oliva reticularis,* and *Lyria (Enaeta) cylleniformis* at Bimini, *Conus columba, Oliva reclusa* and *Oliva fulgurator,* and *Voluta musica* at Aruba, and *Conus puncticulatus, Oliva bewleyi,* and *Voluta virescens* at Cartagena. These species are restricted to their respective areas and this same biogeographical pattern holds for every other group of mollusks and for many groups of marine invertebrates.

Within the Caribbean region, including the Gulf of Mexico and northern Brazil, there are upwards of twenty discrete faunal subregions, each with its own endemic mollusks. If malacologists wished to complete their collections of a certain group of mollusks, say the Muricidae or Conidae, they would need to sample all of these areas, and different depths, to ensure the completeness of the collections. For this reason, the Caribbean region, for its geographically small size when compared to other faunal regions such as the Indo-Pacific, contains the richest tropical marine molluscan fauna in the world. While doing my doctoral research work in northern South America (collecting specimens by working on commercial shrimp boats), I discovered that the Caribbean region is not only made up of a multitude of tiny, discrete faunal subregions, but that some of these subregions are composed of living fossil communities that date from previous geological ages. Although thought to have been extinct for millions of years and known as fossils from around the Caribbean, these living fossil molluscan assemblages are existing in sort of evolutionary stasis, neither becoming extinct nor evolving into new, morphologically-different species complexes. Since the Caribbean region contains a number of archaic, prehistoric, faunal subregions scattered among geologically-younger faunal subregions, it is a perfect example of a new biogeographical pattern that I referred to as geographical heterochrony (PETUCH, 1982).

Geographical heterochrony is a biogeographical pattern found in near-shore environments that represents the persistence, into the Recent, of geologically older faunas. These archaic living faunas, although extinct throughout most of their original ranges, have managed to survive, in a somewhat unaltered state, in geographically small areas that I referred to as relict pockets. Basically, these ancient marine communities are frozen in time and their faunal compositions would easily give a molluscan paleontologist a feeling of scientific *deja-vu.* Two types of relict pockets are known to exist; primary relict pockets, where both the genera, and very often the actual geologically-old species, have survived into the Recent; and secondary relict pockets, where only the archaic genera have survived. In secondary relict pockets, some subsequent speciation has taken place, resulting in the production of new, modern species of supposedly-extinct genera.

Until now, all known relict pockets, both primary and secondary, have been found only in shallow water (0-100 m depths) in or near upwelling systems. Generally, these continuous upwellings of deep, cool, nutrient-rich water appear to result in oceanographic and primary productivity conditions that approximate those of Mio-Pliocene time, and not those of the Recent. In this case, the molluscan assemblages of relict pockets are living in paleoceanographic stasis, having ecologically circumvented the mass extinctions that were caused by environmental changes in shallow water areas around the world during the Pleistocene (STANLEY and CAMPBELL, 1981; STANLEY, 1986; VERMEIJ and PETUCH, 1986). Recent oceanographic explorations along the outer edge of the continental shelf off western Florida have revealed a previously-unrecognized secondary relict pocket. The gastropods of this archaic fauna, unlike those of other known secondary relict pockets, appear to have evolved from their ancestral Neogene fauna by a mechanism that is not related to the persistence of upwellings. The mollusks of this Floridian offshore region will be the first to be described and discussed in this book.

Relict pockets are named by using two criteria; their present geographical location (using some well-known geographical name) and by the relationship of the living fauna of the pocket to ancestral paleoprovinces. The Pliocene and Pleistocene Caribbean contained two very different faunal provinces, each with its own distinctive and unrelated fauna, and was quite different from the Recent Caribbean, which contains a single faunal province. The northern Caribbean province has been named the Caloosahatchian Province (PETUCH, 1982) (for the Caloosahatchee Formation of Florida, whose mollusks typify the fauna of that province), while the southern Caribbean province was referred to as the Gatunian Province (for the Gatun Formation of Panama and Costa Rica, whose mollusks typify the fauna of that province). In naming and defining a relict pocket, then, it is necessary to know the fossil molluscan fauna of an area and its distribution in the past. Using this system of biogeographical nomenclature, I had previously recognized three relict pockets in the Caribbean region; the *Yucatanean Caloosahatchian Relict Pocket* (a secondary relict pocket found along the Campeche Banks and northern side of the Yucatan Peninsula of Mexico), the *Honduran Caloosahatchian Relict Pocket* (a secondary relict pocket found along the northern coast of Honduras, the Bay Islands, and eastern Nicaragua), and the *Venezuelan Gatunian Relict Pocket* (a transitional primary-to-secondary relict pocket found along northern Colombia and Venezuela). A fourth relict pocket has recently been found off Florida and is the subject of the next chapter.

Of the five relatively-unexplored regions covered in this book (Florida, Bahamas, northern Honduras, Colombia and Venezuela, and northern Brazil) three contain relict pockets. For this reason, members of several prehistoric and supposedly-extinct groups such as *Lindapterys* (Muricidae), *Cerithioclava* (Cerithiidae), *Agaronia* (western Atlantic species, Olividae), *Conomitra* (Volutomitridae), *Falsilyria* (Volutidae), *Aphera* (western Atlantic species, Cancellariidae), and *Paraborsonia* (Turridae), have recently been discovered within the geographical boundaries of these relict pockets and can be added to the list of living Caribbean mollusks. It is necessary, then, to view the modern Caribbean molluscan fauna as a patchwork quilt of both discrete spatial and discrete temporal components.

CHAPTER 3

New Discoveries from Florida

Although Florida has been studied more intensely than any other area along the southeastern United States, it still is yielding new malacological surprises on a regular basis. These new discoveries serve as mute testament to the richness of the Floridian molluscan fauna. Florida also offers more varieties of habitats than any other area in the United States and is the boundary region between two large faunal provinces, the Carolinian to the north and in the Gulf of Mexico, and the Caribbean to the south. All of these factors add up to produce one of the most interesting faunal regions in the western Atlantic.

Several areas of the Floridian region have still only been partially studied and sampled. These include the deeper water (200-300 m depth) areas along the edge of the continental shelf in the Gulf of Mexico off western Florida, the deep water (200-250 m depth) coral rubble areas off the coral reefs along southeastern Florida, and the lush Turtle Grass *(Thalassia testudinum)* beds in Florida Bay. The first of these areas, the deep water Gulf of Mexico region, is predominantly an area of sand and coralline algal nodule bottoms. Often, huge growths of sponges will cover over the algal nodule bottom, cementing them into a solid pavement. In these areas, larger basket sponges, gorgonians, and antipatharians grow in profusion. These areas, and the sand and shell gravel patches in between, support an amazing and still poorly-known molluscan fauna. This type of sea bottom houses the first-known relict pocket found along the coastal regions of the United States.

From 1979 to 1984, extensive benthic sampling was undertaken along the outer edge of the western Floridian continental shelf. This research was done by both the marine laboratory of the Florida State Department of Natural Resources and a private environmental consulting firm, Continental Shelf, Inc., Tequesta, Florida. The former sampling project was part of a major research program (the Hourglass Project), designed to explore new harvestable biotic resources and to increase the general knowledge of the deeper water faunas. The latter project involved a survey of the deeper water benthos in conjunction with land-leasing of the outer continental shelf for oil exploration (Shell Oil Co.).

Collections of gastropods from both of these faunal surveys contained many new and interesting species, several of which belonged to genera that were known only from intertidal facies of late Neogene geological formations in the southeastern United States. An analysis of the gastropods of this previously unknown Floridian fauna showed that the outer edge of the western continental shelf contained a new secondary relict pocket. The analysis used the methodology outlined by myself (PETUCH, 1982) and VERMEIJ and PETUCH (1986), which compares the fossil species of an area with the living species and measures the degree of extinction that has taken place. In the case of this relict pocket, many Pliocene intertidal genera are now found to be living in deep, offshore waters. These organisms appear to have adapted to colder water conditions, possibly during low sea levels and colder marine climates at the beginning of the Pleistocene. They have stayed in place, spatially, in the cool, deep areas off western Florida even though the shallow, near-shore environments have now warmed back up to what approximated their original temperature preferences.

Following the nomenclatural scheme of geographical heterochrony that I originally proposed (PETUCH, 1982), I here refer to this new relict fauna as the "Tampan Caloosahatchian Relict Pocket." "Tampan" was chosen because the city of Tampa is the largest and best known city along the western coast of Florida and is a good geographical marker for this new relict pocket. Some of the relict genera that have survived within the Tampan Relict Pocket are *Heilprinia* (Fasciolariidae), *Muricantharus* (Buccinidae), *Busycoarctum* (Melongenidae), and *Myurella* (Terebridae). The new species of these genera are described later in this chapter. Besides containing unusual genera and species, the structure of the molluscan communities off western Florida are also bizarre. As seen in the Hourglass collections, at 150-200 m depth due west of Tampa, the slit shell *Perotrochus amabilis* (Bayer) coexists, in relative abundance, with the busycon whelk *Busycon (Busycoarctum) lyonsi* Petuch (described here). This is the only ecosystem known in the western Atlantic where a very deep water group (*Perotrochus* slit shells) lives side-by-side with a basically intertidal or shallow water group (*Busycon* whelks). Also found living within this odd assemblage is the tulip shell *Fasciolaria (Cinctura) bullisi* Lyons (another basically shallow water group) (Plate 7, Figure 1), the auger shell *Terebra (Strioterebrum) onslowensis* Petuch (originally named from North Carolina) (Plate 6, Figures 15, 16), and the spindle shell *Fusinus stegeri* Lyons (Plate 6, Figure 3). The following is a listing of new taxa from this area.

New Species From The Tampan Relict Pocket

Turritella lindae new species
Cyphoma lindae new species
Chicoreus rachelcarsonae new species
Favartia lindae new species
Murexiella kalafuti new species
Heilprinia lindae new species
Busycon (Busycoarctum) lyonsi new species
Muricantharus harasewychi new species
Cancellaria richardpetiti new species
Conus aureonimbosus new species
Terebra (Myurella) lindae new species
Polystira sunderlandi new species
Splendrillia sunderlandi new species

Family **Turritellidae**
Genus *Turritella* Lamarck, 1799
Subgenus *Torcula* Gray, 1847

Turritella (Torcula) lindae new species (Plate 1, Figures 1, 2)

Material Examined: HOLOTYPE - Length 24 mm, width 7 mm USNM 859802, dredged from 150 m depth, 50 km south of Apalachicola, Florida, 1982; PARATYPE - Length 18 mm, same locality, depth, and data as holotype, Petuch collection.

Description: Shell elongated, average size and shape for subgenus *Torcula;* whorls ornamented with 5 large cords with numerous fine threads between; anteriormost 3 cords larger, intersected by large vertical ribs; 12 ribs per whorl on holotype; suture slightly keeled; protoconch comprising 3 whorls; protoconch and early whorls bright purple in color; shell color white with scattered fine purple and brown dots.

Etymology: Named for my wife, Linda J. Petuch.

Discussion: *Turritella lindae,* is closest to the similar-sized *T. acropora* Dall, 1881, but differs in having a much coarser, beaded sculpture pattern and by lacking the brown or orange flammuled color pattern. Both *T. acropora* and *T. lindae* occur together off northwestern Florida. The new species is similar to *Turritella ochlockoneensis* Mansfield from the Pliocene Jackson Bluff and Buckingham Formations of Florida, and may be a descendant of that lineage.

Family **Ovulidae**
Genus *Cyphoma* Röding, 1798

Cyphoma lindae new species (Plate 1, Figures 3,4)

Material Examined: HOLOTYPE - Length 18 mm, width 5 mm, dredged from 150 m depth, 50 km south of Apalachicola, Florida, 1982, USNM 859806; PARATYPE - length 15 mm, same depth, location, and date as holotype, Robert Pace Collection.

Description: Shell small for genus, inflated, thin and fragile, roughly diamond-shaped; lateral calluses thin, poorly-developed; dorsal transverse ridge strong, well-developed; aperture wide, flaring; anterior end of columella with fossula; color brownish purple, becoming paler on extremities and on transverse ridge; interior of aperture brownish-purple; mantle of animal (from preserved specimen) transparent, peppered with abundant, tiny, black speckles and larger black dots and blotches; margin of foot marked with numerous vertical black lines.

Etymology: Named for my wife, Linda J. Petuch.

Discussion: This new deep water *Cyphoma* is morphologically closest to the shallow water Floridian *Cyphoma macgintyi* Pilsbry, but differs in a number of ways; by having a smaller, stumpier, more truncated shell, by being a much more fragile species, by having a different shell color (purple-brown as opposed to lilac or pink), and by having a different colored animal. Antipatharian octocorallians were commonly dredged along with the type specimens of *Cyphoma lindae,* and the new species may be a predator on *Antipathes* and *Spirapathes* species. *Cyphoma lindae* is also similar to *C. macumba* Petuch (Plate 15, Figures 17, 18) from Brazil, but is smaller and lacks the prominent marginal projection of *C. macumba.*

Family **Muricidae**
Genus *Chicoreus* Montfort, 1810

Chicoreus rachelcarsonae new species (Plate 1, Figures 9, 10)

Material Examined: HOLOTYPE - Length 43 mm, width 25 mm (with spines), trawled from 150 m depth, 50 km south of Apalachicola, Florida, 1982, USNM 859811; PARATYPES - Length 23 mm, dredged from 70 m depth, 50 km due north of Marquesas Keys, Florida Keys, ANSP 354768; length 41 mm, collection of Robert Pace, Miami, Florida.

Description: Shell delicate, very frondose, with 3 varices per whorl; one large knob along shoulder between varices; varix with 5 spines; spines often frondose; siphonal canal with 3 large spines; spine on shoulder largest; spire high and elevated; shell color pink or pale salmon-orange overlaid with numerous fine, pale tan spiral threads; spines often darker pinkish red or orange-red; protoconch and early whorls bright orange-red; siphonal canal completely closed.

Etymology: Named for the late Rachel L. Carson, marine biologist, naturalist, and well-known author, whose books *The Sea Around Us, The Edge of the Sea, Under the Sea Wind,* and *Silent*

Spring, were an inspiration to me as a young student of the natural history of the sea.

Discussion: This deep water, beautiful little species is presently known only from the outer edge of the continental shelf off western Florida. It is closest to the common, shallow water *Chicoreus dilectus* (A. Adams) and *C. florifer* (Reeve), but differs from both of those species in being a more delicate, spinose species, with a larger aperture proportionally, by being consistently colored in pinks, oranges, and reds, and in consistently having only five spines per varix instead of six. *Chicoreus rachelcarsonae,* in having pinkish-red varical spines, somewhat resembles the deep water Philippine *C. artemis* Radwin and D'Attilio.

Genus *Favartia* Jousseaume, 1880

Favartia lindae new species (Plate 1, Figures 5, 6)

Material Examined: HOLOTYPE - Length 11 mm, width 7 mm, dredged from 150 m depth, 50 km south of Apalachicola, Florida, 1982, USNM 859805; PARATYPE - length 14 mm, same locality, depth, and date as holotype, in Petuch collection.

Description: Shell small for genus, thin, delicate, elongated, fusiform; 5 varices per whorl; 6 major cords around body whorl, with 1 minor cord in-between; cords frilly, fimbriated; varices coarsely ribbed, cancellated; siphonal canal with 2 large, fimbriated cords; siphonal cords end in 2 large spines on each varix on siphonal canal; aperture large in proportion to shell size; shell color pale tan with varices being darker tan-brown; spire elevated; protoconch and early whorls brown.

Etymology: Named for my wife, Linda J. Petuch.

Discussion: This tiny, high-spire species is quite different from any known western Atlantic *Favartia.* It is possible to confuse *F. lindae* with the shallow water *F. cellulosa* (Conrad), but the new deep water species differs in being a smaller, more delicate shell with thinner, less-developed varices, and by having a higher, more protracted spire. *Favartia lindae* also differs from *F. cellulosa* by being a much more colorful shell, with chocolate-brown varices and early whorls. In this last character, *F. lindae* somewhat resembles the obese, compactly-built, and thickened *F. nucea* (Mörch) from shallow water areas in the southern Caribbean.

Genus *Murexiella* Clench and Farfante, 1945

Murexiella kalafuti new species (Plate 1, Figures 7, 8)

Material Examined: HOLOTYPE - Length 16 mm, width 10 mm (with spines), dredged from 150 m depth, 50 km south of Apalachicola, Florida, 1982, USNM 859810; PARATYPES - 5 specimens in Kalafut collection, Key West, Florida, from lobster traps set in 50 m water off Key West, Florida Keys; 10 specimens in Pace collection, Miami, Florida, from lobster traps off Key West.

Description: Shell small for genus, fusiform, with elevated spire; 6 varices per whorl; shoulder sharp-angled; body whorl with 5 large cords; cords end in thick, blunt spines at intersection with varix; cords minutely fimbriated; varices heavily fimbriated; shell grayish-white with 2 wide, dark brown bands, one around mid-body and one around siphonal canal; suture marked with row of intermittent dark brown patches; siphonal canal with 2 large cords, producing 2 large spines at intersection with varix; aperture ovate, large in proportion to shell size; dark mid-body band showing through in aperture.

Etymology: Named for Theodore (Ted) Kalafut of Key West, Florida, who first collected the new species from lobster traps in Key West.

Discussion: In the western Atlantic, *Murexiella kalafuti* is closest to the shallow water *M. glypta* (M. Smith), but differs in being a stumpier shell with a lower spire, by having fewer foliations and fimbriations on the body whorl, by having only two spines on the siphonal canal instead of three, and by having only six varices per whorl instead of eight. The banded color pattern of *M. kalafuti* also readily differentiates the new species from *M. glypta. Murexiella kalafuti* is most similar to *M. vittata* (Broderip) from the Panamic Province, especially in having a banded color pattern, and probably represents the Atlantic analogue species. *Murexiella glypta* is well-illustrated by FAIR (1976), and this can be used for making comparisons with *M. kalafuti.*

Family **Fasciolariidae**
Subfamily **Fusininae**
Genus *Heilprinia* Grabau, 1904

Heilprinia lindae new species (Plate 2, Figures 1, 2)

Material Examined: HOLOTYPE - Length 143 mm, width 44 mm, dredged from 150 m depth, 50 km south of Apalachicola, Florida, 1982, USNM 859908; PARATYPE - Length 120 mm, same depth, locality, and date as the holotype, Petuch collection.

Description: Shell very elongated, with spire greatly protracted and siphonal canal extended; aperture oval, flaring around edge, with raised parietal shield; shoulder rounded; body whorl completely rounded, ornamented with 19 smooth, spiral cords; numerous thin, spiral threads between cords; siphonal canal ornamented with 30 large spiral cords; numerous spiral threads between siphonal cords; interior of lip with 19 raised cords that correspond to cords on body whorl; first 7 whorls ornamented with large, knoblike axial ribs; edge of suture marked with thin but deeply-incised, canaliculate sulcus; columellar area of parietal shield ornamented with numerous ribs and crenulations; shell color pale pinkish-orange, overlaid with thin, pale orange bands that correspond to the spiral cords on body whorl and siphonal canal; paratype pure white, with early whorls being pale orange; periostracum thick, fuzzy, adherent.

Etymology: Named for my wife, Linda J. Petuch.

Discussion: This large new species is the second known *Heilprinia* from the Gulf of Mexico. *Heilprinia lindae* can be easily separated from the well-known Turnip Spindle *H. timessus* (Dall), by a number of ways; by having a much higher, protracted spire, by being a much more slender shell, by lacking the prominent axial ribs on the body whorl that characterize *H. timessus,* by having finer spiral sculpturing, and by having the deeply-incised, canaliculate sulcus around the suture. *Heilprinia lindae* is also a much more delicately-colored shell, ranging from white to pastel pink and pinkish-orange, in contrast to the orange, tan, and orange-brown shells of *H. timessus. Heilprinia dalli* (Mansfield) from the Pliocene Jackson Bluff and Buckingham Formations of Florida is a similar, although smaller, shell and may be a direct ancestor of *H. lindae.*

Family **Melongenidae**
Subfamily **Busyconinae**
Genus *Busycon* Röding, 1798
Subgenus *Busycoarctum* Hollister, 1958

Busycon (Busycoarctum) lyonsi new species (Plate 1, Figures 11, 12, 13)

Material Examined: HOLOTYPE - Length 132 mm, width 69 mm, dredged from 400 ft depth, 70 miles due west of Clearwater, Florida, by HOURGLASS research project, Florida State Dept. of

Natural Resources Marine Laboratory, 1979, USNM 859803; PARATYPE - Length 135 mm, same depth, locality, and date as holotype, USNM 859804.

Description: Shell small for genus, very thin and fragile, with body whorl inflated; siphonal canal long, very narrow in proportion to body whorl; spire low; shoulder carinated, with 15-18 sharp, spikelike knobs on last whorl; aperture wide, flaring; columella with one large, slanting fold; interior of lip with numerous ribs that extend back into interior of aperture; protoconch very large in proportion to shell size, composed of 2 whorls, orange in color; shell color bright pinkish-orange overlaid with 3 wide bands of darker orange, one below shoulder, one below mid-body, and one around anterior end of siphonal canal; shell marked with numerous prominent dark orange-brown, thin, longitudinal flammules running entire length of shell; spire whorls with regularly-spaced dark orange-brown radiating flammules; interior of aperture bright orange; periostracum thin, adherent.

Etymology: Named for William Lyons, marine biologist, Florida State Department of Natural Resources Marine Laboratory, St. Petersburg, Florida, who kindly donated the type specimens for study.

Discussion: *Busycon lyonsi* is closest to *B. candelabrum* (Lamarck) from the western Gulf of Mexico. The new Florida species differs from its Mexican relative by being a smaller, more fragile shell, by having a lower spire, by having a narrower and longer siphonal canal, and by being a pinkish-orange shell with a bright orange aperture. ABBOTT (1974, Color Plate 12, figure 2442) illustrates, in color, a classic cream-colored and yellow *B. candelabrum* that can be used for comparison with the orange *B. lyonsi.* Ecologically, *Busycon lyonsi* also differs from *B. candelabrum.* The new Florida species occurs in much deeper water (in excess of 400 ft.) than does its Mexican cogener, which is collected by shrimp boats from 60-100 ft. *Busycon lyonsi,* then, represents the deepest-dwelling busyconine species. Mr. Lyons (pers. comm.) states that *Busycon lyonsi* lives together with *Chicoreus rachelcarsonae* Petuch, *Fasciolaria bullisi* Lyons, and the slit shell *Pleurotomaria (Perotrochus) amabilis* Bayer, making this molluscan assemblage one of the strangest in the world.

Family **Buccinidae**

Genus *Muricantharus* Olsson, 1971

Muricantharus harasewychi new species (Plate 2, Figures 3, 4)

Material Examined: HOLOTYPE - Length 25 mm, width 15 mm, dredged from 150 m depth, 50 km south of Apalachicola, Florida, 1982, USNM 859801; PARATYPE - length 25 mm, trawled from 70 m depth, 50 km due north of the Marquesas Keys, Florida Keys, ANSP 354769.

Description: Shell thin, inflated, with elevated spire; 8 strong varices per whorl; body whorl ornamented with 4 large spiral cords; numerous fine threads between large cords; 4 large knobs per varix, produced by intersection of varix with spiral cord; siphonal canal short, stumpy, ornamented with numerous spiral threads; interior of aperture with large, well-developed plications on inner side of lip; suturs incised, indented, producing scalariform effect; shell color bright yellow-orange, orange-red, or salmon-pink; paratype (ANSP) with anterior half of body whorl bright pinkish-orange, posterior half bright yellow, producing two-toned color pattern; aperture and columella white.

Etymology: Named for Dr. M. G. Harasewych, Division of Mollusks, NMNH, Smithsonian Institution.

Discussion: The is the first *Muricantharus* species to be found in the Atlantic Ocean. Previously (OLSSON, 1971, p. 61), only one species, *M. panamicus* (Hertlein and Strong), was recognized. KEEN (1971, pp. 558-560) recognizes three closely-related species in the eastern Pacific, *M. pan-*

amicus, M. rehderi (Berry), and *M. shaskyi* (Berry). Of these three, *M. harasewychi* is closest to *M. shaskyi*, but differs in being a more slender shell, by having only four knobs on each varix, and by having a sharper-angled shoulder. *Muricantharus harasewychi* can be differentiated from all possible morphs of the shallow water *Cantharus multangulus* (Philippi) by having knobbed varices.

Family **Cancellariidae**
Genus *Cancellaria* Lamarck, 1799

Cancellaria richardpetiti new species (Plate 2, Figures 5, 6)

Material Examined: HOLOTYPE - Length 31 mm, width 16 mm, dredged from 150 m depth, 50 km south of Apalachicola, Florida, 1982, USNM 859809.

Description: Shell thin, fragile, composed of 6 whorls; spire protracted, body whorl elongated; body whorl and spire whorls ornamented with longitudinal ribs; 32 ribs on body whorl, running from suture to anterior tip; ribs on spire whorls more prominent, producing varix-like appearance; longitudinal ribs intersected by raised, spiral cords; large bead produced where rib and cord intersect, forming general appearance of longitudinal lines of beads; protoconch very large; aperture narrow; columella with 3 folds, posteriormost being the largest; color white, with 3 wide, uninterrupted bands of orange-brown around body whorl, one along shoulder, one anterior of mid-body, and one around siphonal canal; spire whorls solid orange-brown with thin white band along suture; interior of aperture white; 3-whorled protoconch dark brown; interior of aperture white.

Etymology: Named for Richard E. Petit, Myrtle Beach, South Carolina, in recognition of his many contributions to the systematics of the Cancellariidae.

Discussion: The deep water *Cancellaria richardpetiti* is similar to the wide-ranging Caribbean, shallow water *C. reticulata* (Linnaeus), but differs in being a smaller, much more slender shell, by having sculpturing composed of longitudinal rows of beads in a riblike pattern, by having a larger protoconch composed of three whorls, by having three columellar folds, and by having a color pattern made up of three uninterrupted brown bands. ABBOTT (1974, Color Plate 10, figure 2683) illustrates, in color, a specimen of *C. reticulata*, showing its characteristic color pattern of large brown checkers and blotches, which is so different from *C. richardpetiti*.

Family **Conidae**
Genus *Conus* Linnaeus, 1758

Conus aureonimbosus new species (Plate 2, Figures 7, 8)

Material Examined: HOLOTYPE - Length 27 mm, width 13 mm, dredged from 150 m depth, 50 km south of Apalachicola, Florida, 1982, USNM 859812; PARATYPE - Length 26 mm, trawled from 70 m depth, 50 km due north of the Marquesas Keys, Florida Keys, ANSP A9145 (preserved in alcohol collection).

Description: Shell thin, fragile, slender and elongated; body whorl highly polished; numerous fine spiral cords around anterior end; shoulder sharp, obsoletely coronated with low undulations and rounded bumps; spire low; protoconch needle-like, protracted, projecting above spire; shell color pale cream-yellow overlaid with large, amorphous, flammules of bright golden-yellow; mid-body with white band containing rows of pale tan dots and dashes; spire whorls white with dark orange and tan flammules; protoconch yellow; interior of aperture white; anterior tip of shell yellow; periostracum thin, yellow, transparent.

Etymology: In reference to the bright yellow cloudings on the body whorl that are characteristic of this new species.

Discussion: *Conus aureonimbosus* is closest to *C. amphiurgus* Dall (= *C. juliae* Clench) from the Gulf of Mexico and Carolinian Province, but differs in being a much smaller, more elongated and slender shell, by having a coronated spire, and by having a protracted, mammilate protoconch. By having an obsoletely coronated spire, *C. aureonimbosus* may actually be closer to the *C. cardinalis* complex from the Caribbean. If this is the case, then *C. aureonimbosus* represents the deepest-dwelling member of that normally reef-dwelling species group. The brilliant yellow patches on this new species certainly sets it apart from all other western Atlantic cone shells. *Conus aureonimbosus* and *C. amphiurgus* are compared on Plate 4.

Family **Terebridae**
Genus *Terebra* Bruguière, 1789
Subgenus *Myurella* Hinds, 1845

Terebra (Myurella) lindae new species (Plate 2, Figures 11, 12)

Material Examined: HOLOTYPE - Length 63 mm, width 10 mm, dredged from 150 m depth, 50 km south of Apalachicola, Florida, 1982, USNM 859807; PARATYPES - length 68 mm (tip broken), same depth, locality, and date as holotype, Gene Everson collection; length 65 mm, same locality, Petuch collection.

Description: Shell very elongated, slender; suture bordered by row of very large, rounded beads; row of smaller beads below (anterior to) row of large beads; rest of whorl smooth, shiny; color orange-tan with two rows of larger, reddish-brown spots on each whorl, one row along suture, often between subsutural beads; columella and siphonal canal bright yellow; interior of aperture orange-tan.

Etymology: Named for my wife, Linda J. Petuch.

Discussion: *Terebra lindae* is closest to the widespread *Terebra floridana* Dall, which ranges from South Carolina to Brazil. The new Florida species differs from its widespread Caribbean relative by being a much more elongated and slender species and by having larger subsutural beads. The most obvious difference between the two species is seen in the color; *T. floridana* is a uniformly unicolored yellow or yellowish-white shell without any spots, while *T. lindae* is a bright orange-tan shell with two rows of large, dark brown or reddish-brown spots and patches. By having a spotted shell, *T. lindae* somewhat resembles a pale, very slender *T. taurina* Lightfoot. *Terebra lindae* is also a much smoother shell than *T. floridana,* virtually lacking any sculpturing below the subsutural beads.

Family **Turridae**
Subfamily **Clavinae**
Genus *Splendrillia* Hedley, 1922

Splendrillia sunderlandi new species (Plate 2, Figures 9, 10)

Material Examined: HOLOTYPE - Length 40 mm, width 14 mm, dredged from 150 m depth, 50 km south of Apalachicola, Florida, 1982, USNM 859800; PARATYPE - Length 20 mm (juvenile), Kevan Sunderland collection.

Description: Shell large for genus, elongated; body whorl and spire whorls rounded, without obvious shoulder; shell smooth, silky, without spiral grooves, threads, or other sculpture; body whorl with large, wide, swollen axial hump that is roughly 3 ribs in width; columella wide, well-developed; shell color pure white with single, wide, wine red band around mid-body; spire whorls with red band running along suture; interior of aperture white, with red band showing through from exterior.

Etymology: Named for Kevan Sunderland, Fort Lauderdale, Florida, in recognition of his explorations and important discoveries in the Caribbean malacofauna.

Discussion: *Splendrillia sunderlandi* has been collected in the past off western Florida, but always was identified as *Splendrillia janetae* (Bartsch), a closely-related but different species from farther south in the Caribbean. *Splendrillia sunderlandi* differs from *S. janetae* in being a larger shell with much more rounded, sloping shoulders, by lacking the coarse sculpturing of spiral threads and grooves seen on *S. janetae,* and by having a thinner red band around the mid-body. The large hump on the dorsum of the last whorl of *S. sunderlandi* is similar to those seen on fossil species of the genus *Cymatosyrinx* Dall, 1889, and the new species may actually belong to that genus and not *Splendrillia.*

Genus *Polystira* Woodring, 1928

Polystira sunderlandi new species (Plate 6, Figures 13, 14)

Material Examined: HOLOTYPE - Length 28 mm, width 8 mm, dredged from 150 m depth, 50 km south of Apalachicola, Florida, 1982, USNM 859896; PARATYPE - 1 specimen, Kevan Sunderland collection, same locality and depth as holotype.

Description: Shell small for genus, fusiform, elongate, with high, elevated spire; 6 raised, bladelike, spiral cords on body whorl; spire whorls with 3 cords; cords all equally-spaced and equal-sized; siphonal canal long, straight, with numerous small spiral cords and spiral threads; anal notch narrow, deep, aligned with second cord; shell color light brown with single narrow white band that corresponds to second cord from suture; siphonal canal tan with dark brown anterior tip; interior of aperture tan.

Etymology: Named for Mr. Kevan Sunderland, Ft. Lauderdale, Florida.

Discussion: *Polystira sunderlandi* is similar to *P. florencae* Bartsch from Puerto Rico, both in size and basic color pattern, but differs in being a more elongated with fewer cords on the body whorl, and in having a narrower white band around the shoulder. In *P. florencae,* the white shoulder band and cord are wider that the other body whorl cords, while in *P. sunderlandi,* the white shoulder cord is the same size as the other cords. The brown subsutural cord of *P. florencae* is more indented than the wide white shoulder cord, producing a slope from the suture to the white shoulder cord, and giving the shell and spire a tabulated appearance. In *P. sunderlandi,* the brown subsutural cord has the same circumference as the white shoulder cord, giving the shell and spire a tapered appearance. *Polystira sunderlandi* is known only from the northeastern corner of the Gulf of Mexico.

Other New Floridian Faunas

Beside the rich and interesting fauna of the Tampan Relict Pocket, four other subregions of the Floridian area are still yielding new species. These are the deep, coral rubble environments off Palm Beach and Broward Counties, the intertidal beach rock areas from Fort Pierce south to Palm Beach, the deep water (200-300 m depth), soft bottom areas in the Straits of Florida, and the Turtle Grass communities of Florida Bay.

The continental shelf off Palm Beach and Broward Counties is very narrow, and relatively deep water conditions exist in the near-shore area. The Gulf Stream also comes close to shore along this stretch of coast, and bathes the shelf area in perpetual tropical water. Because of this, large coral patch reefs grow along the edge of the shelf, and their dead fragments tumble down the steep-to shelf as a series of giant coral talus slopes. SCUBA divers have recently explored these coral rubble areas (30-80 m depth), and have dug into the talus fauna. The rare cone shell, *Conus granulatus* Linnaeus, was found to be plentiful in these rubble areas, living buried as much as one meter within the coral fragments. Another new reef cone of the *Conus jucundus* Sowerby species complex was also found in this area, and is described here. The rare and poorly-known miter, *Mitra pallida* Usticke (Plate 7, Figure 9), was also found to coexist, deep within the rubble, with *Conus granulatus.* New species of the muricaceans *Coralliophila, Muricopsis,* and *Dermomurex,* were also found in this cryptic fauna and are described in this chapter.

Dredging in deep water in the Straits of Florida, from Palm Beach to the Keys, has also yielded large numbers of new and seldom-seen mollusks. Some inspired amateurs, such as Dr. Marvin Glickstein of North Palm Beach, have built elaborate dredging systems which they run from their own boats. In the case of Dr. Glickstein, the sampling gear is comparable to that of many professional oceanographic research vessels. Because of the dedication of these collectors, several new and important gastropods have been trawled off Palm Beach and Broward Counties, including the rare *Coralliophila aedonis* (Watson) (Plate 4, Figures 8, 9), *Cymatium pharcidum* (Dall) (Plate 4, Figures 4, 5), and *Conus flamingo* Petuch (Plate 6, Figures 10, 11, 12). Deeper water dredging in the Straits of Florida between the Bahamas and Florida and off the Florida Keys have brought to light a number of rarely-seen species such as the turbinellids *Fulgurofusus aurora* (Bayer), *Fulgurofusus electra* (Bayer), and *Fulgurofusus merope* (Bayer) (shown on Plate 8), the muricid *Pterynotus guesti* Harasewych and Jensen (Plate 8, Figures 1, 2), and the fasciolariids *Latirus macmurrayi* Clench and Aguayo (Plate 6, Figures 5, 6) and *Latirus varai* Bullock (Plate 8, Figure 13). Farther north, along the coasts of South Carolina and Georgia, a large and interesting turbinellid was recently described as *Fulgurofusus timor* (Harasewych) (Plate 4, Figures 15, 16, 17). This beautiful species may eventually be found along the coast of Florida.

The Turtle Grass beds of Florida Bay are amongst the lushest in the Caribbean region. Within the confines of the Everglades National Park, these grass beds contain a rich and partially endemic molluscan fauna. Commonly encountered crawling on the grass blades in the central region of the Bay are a number of new, and apparently endemic, mollusks including a new *Zafrona* dove shell, a new *Haminoea* bubble shell, both of which are described later in this chapter, and a still un-named small green chiton. Also found in these *Thalassia* beds are a delicate, spinose form of *Eupleura sulcidentata* Dall, a large, elongated, spineless form of *Melongena corona* (Gmelin), and a new subspecies of the scallop *Argopecten irradians* (described here). This unusual and intriguing Floridian ecosystem is discussed in detail under the individual species descriptions later in this chapter. As can be seen from this preliminary research, the Florida Bay component of Everglades National Park is still in need of more in-depth faunal surveys.

The beachrock areas of southeastern Florida, which run from near Cape Canaveral south to the Florida Keys, are remnants of dead coral reefs that were exposed when sea levels dropped to their present-day level about 75,000 years ago. They were deposited when sea levels were about 5 meters higher during the Sangamon Interglacial Stage of the Pleistocene, about 150,000 years ago. Being in an area that has tropical temperatures during the summer but is exposed to freezing temperatures in the winter, the intertidal mollusks of these beachrock environments live in harsher conditions than do related species in similar environments farther south in the Caribbean. Because of these unique ecological conditions and genetic isolation during the Pleistocene, several endemic rocky intertidal species have evolved along the southeastern coast of Florida. Among these endemics are a new *Cerithium* species and a new *Modulus,* both of which are described later in this chapter.

Included on the list of important molluscan discoveries from Florida are two which are particularly interesting from a biogeographical viewpoint. These include the first-known western Atlantic species of the Indo-Pacific and Panamic coralliophilid genus *Quoyula,* and the first-known species of the Panamic capulid genus *Cyclotheca,* which was previously known from a single species from the western coast of Nicaragua. These two interesting new species, and new records for the western Atlantic, are described and discussed later in the chapter.

New Species from Southeastern Florida, the Florida Keys, and Florida Bay

Architectonica sunderlandi new species
Cyclotheca pacei new species
Cerithium lindae new species
Modulus kaicherae new species
Modulus lindae new species
Modulus pacei new species
Dermomurex (Trialatella) glicksteini new species
Murexiella taylorae new species
Lindapterys vokesae new species (fossil)
Muricopsis richardbinghami new species
Coralliophila pacei new species
Quoyula kalafuti new species
Zafrona taylorae new species
Oliva (Strephona) bifasciata sunderlandi new subspecies
Conus binghamae new species
Conus floridanus patglicksteinae new subspecies
Conus glicksteini new species
Conus harasewychi new species
Haminoea taylorae new species
Argopecten irradians taylorae new species

Family **Architectonicidae**
Genus *Architectonica* Röding, 1798

Architectonica sunderlandi new species (Plate 10, Figures 1, 2, 3, 4)

Material Examined: HOLOTYPE - Height 10 mm, diameter 20 mm, trawled by deep water shrimpers from 250 m depth off Key West, Florida Keys, Florida, 1982; PARATYPE - diameter 31 mm, same depth and locality as holotype, Kevan Sunderland collection.

Description: Shell discoidal, flattened, with keeled periphery; dorsum of shell sculptured with 6 low cords per whorl; subsutural cord and 2 cords along periphery larger than central 3 cords; cords intersected by radiating axial grooves; base of shell smooth, without cords; periphery of umbilicus marked with radiating riblets; color orange with wide reddish-orange flammules on subsutural cord and two peripheral cords; base of shell white with pale orange area around periphery of umbilicus; early whorls purple; protoconch and postnuclear whorls magenta.

Etymology: Named for Mr. Kevan Sunderland, Fort Lauderdale, Florida, who collected the type lot.

Discussion: *Architectonica sunderlandi* differs from the widespread, shallow water *A. nobilis* Röding in being a smaller flatter shell, with a much less developed, smoother sculpture pattern. *Architec-*

tonica nobilis is cream colored with numerous large brown spots, while *A. sunderlandi* is orange with reddish-orange flammules arranged axially around the periphery. The base of *A. nobilis* is heavily sculptured with beaded cords, while the base of *A. sunderlandi* is smooth. *Architectonica sunderlandi* could also be confused with *A. peracuta* (Dall), but differs in having a higher spire, less-developed peripheral keel, and in having orange flammules around the periphery. To date, *A. sunderlandi* has only been collected by deep water shrimpers (royal red shrimp) off Key West, Florida.

Family **Capulidae**
Genus *Cyclotheca* Stearns, 1890

Cyclotheca pacei new species (Plate 4, Figures 6, 7)

Material Examined: HOLOTYPE - Length 9 mm, width (height) 4 mm, on the back of the scallop, *Chlamys sentis,* under rocks in 1 m depth, beneath the MacArthur Causeway Bridge, Biscayne Bay, Miami, Florida, 1979, USNM 859816; PARATYPES - length 7 mm, same locality and depth as holotype, USNM 859817; 20 specimens, same locality and depth as holotype, Robert Pace collection and 3 specimens Kevan Sunderland collection.

Description: Shell small, with only 1 1/2 - 2 coils, with body whorl rapidly enlarging to produce cup-shaped appearance; shell sculptured with 3 large prominent ribs, one around shoulder and two around base; shoulder rib widely separated from two anterior ribs; spire flattened; shell color pale grayish-white, overlaid with mottled dark gray, smooth, highly adherent periostracum; interior of aperture white; ribs white or mottled gray; ribs of body whorl project beyond margin of lip, producing 3-pronged, serrated edge.

Etymology: Named for Mr. Robert Pace of Miami, Florida, who collected the type lot.

Discussion: *Cyclotheca pacei* is the first member of this unusual capulid genus to be found in the Atlantic Ocean. The genus *Cyclotheca* was previously known from a single rare species, *C. corrugata* Stearns, which was collected along the western coast of Nicaragua in the late 1800's and has never been seen again. *Cyclotheca pacei* differs from the Panamic *C. corrugata* in being a larger shell with fewer ribs (3 as opposed to 8), and in having a much lower spire. The ecology of the genus *Cyclotheca* is as interesting as its distribution. To date, *Cyclotheca pacei* has only been found in one location; under rocks in shallow water beneath and near the MacArthur Causeway Bridge between Miami and Miami Beach, at the north end of Biscayne Bay. Here, it lives attached to the outsides of the valves of several species of sessile, epibyssate scallops that attach to the bottom of sponge-covered rocks. Mr. Pace reports (in litt.) that the new species live on *Chlamys sentis, Chlamys mildredae,* and *Chlamys imbricata,* often attaching to the auricles or near the edge of the lip. Where attached, *Cyclotheca pacei* puts down a dark-colored layer of periostracum-like material (Plate 4, Figure 3). There is often a clean area surrounding the dark attachment spot. According to Mr. Pace, *C. pacei* seems to prefer *Chlamys imbricata* as a substrate, since he found some specimens with as many as 5 *C. pacei* attached to each individual.

Family **Cerithiidae**
Genus *Cerithium* Bruguière, 1789

Cerithium lindae new species (Plate 3, Figures 8, 9, 10)

Material Examined: HOLOTYPE - Length 11 mm, on rocks exposed at low tide, rock reef on northeastern side of Peanut Island, Palm Beach Inlet of Lake Worth, West Palm Beach, Florida,

1983, USNM 859820; PARATYPES - 12 specimens, same locality and date as holotype, USNM 858821; 5 specimens, same locality and date as holotype, Petuch collection; 20 specimens, Pace collection; 10 specimens, Sunderland collection.

Description: Shell small for genus, very elongated, with rounded whorls; whorls without varices, circular in cross-section; cylindrical in shape; siphonal canal short, open, barely developed; outer lip flaring, thickened, producing single varix (on adult specimens); aperture round; body whorl sculptured with 9-12 large beaded cords; subsutural beaded cord often largest, separated from other body whorl cords by narrow gap without sculpturing; shell color grayish-white with beads on spiral cords being black; inner edge of lip with evenly-spaced, black dots; cords often with one white bead between two black beads, producing a tessellated appearance; interior of aperture white.

Etymology: Named for my wife, Linda J. Petuch.

Discussion: This interesting little species appears to be endemic to the southeastern coast of Florida, where it occurs on beachrock shorelines from near Fort Pierce (Dr. Kerry CLARK, FIT, pers. comm.), to Blowing Rocks near Jupiter (William LYONS, pers. comm), and south to central Palm Beach County. Along its range, *Cerithium lindae* lives together with the similar-looking *Batillaria minima* (Gmelin), which may explain why it has been overlooked for all these years. Some workers, such as ABBOTT (1974, p. 105), have confused this Florida endemic species with the Bahamian endemic, *Cerithium biminiense* Pilsbry and McGinty. The Bahamian species, which lives in the highest tide pools along with the nerite *Puperita pupa* and the periwinkle *Littorina mespillum,* is a smooth, stubby, stocky species, itself resembling a periwinkle, with a zebra pattern of pale blue and black zig-zag flammules.

Cerithium lindae is closest to *C. lutosum* Menke, a widespread Caribbean species, but differs in being a much more slender, elongated shell without varices on the body whorl, in having finer sculpturing with smaller beading, in having the characteristic gap between the subsutural beaded cord and the rest of the close-packed beaded cords on the body whorl, and having the extremely flaring outer lip on adults that produces the characteristic single, final varix. The color patterns of *C. lindae* and *C. lutosum* also differ; *C. lindae* is basically a white shell overlaid with rows of black beads, while *C. lutosum* is a much darker-colored shell, being uniformly dark brown or reddish brown with scattered white flammules and speckling.

Cerithium lindae disappears south of Palm Beach County, and has not been taken on beach rock outcrops along Biscayne Bay or the Florida Keys. Here the new species is replaced by *C. lutosum,* which also co-exists with *Batillaria minima. Cerithium lindae* may have evolved from a population of *C. lutosum* that was isolated along the Florida coast by late Pleistocene sea level drops. As a high intertidal species, *C. lindae* also probably evolved a tolerance to the cold and freezing temperatures that occur commonly during the winter months along the coastlines north of the Florida Keys. This temperature barrier keeps the tropical, stenothermal *C. lutosum* restricted to the Florida Keys and Florida Bay. The eurythermal, more temperature-tolerant *Batillaria minima* can occur throughout the region; occuring with *C. lindae* in the northern area, and *C. lutosum* in the south.

Family **Modulidae**
Genus *Modulus* Gray, 1842

Modulus kaicherae new species (Plate 3, Figures 3, 4)

Material Examined: HOLOTYPE - Diameter 14 mm, height 11 mm, trawled from 70 m depth off Palm Beach Island, Palm Beach, Florida, 1983, USNM 859824; PARATYPE - 1 specimen, same depth and locality as holotype, Glickstein collection.

Description: Shell flattened, discoidal, with low spire; shoulder and whorls with 8 large, pointed knobs; pointed knobs project beyond shell periphery, giving shell appearance of small *Arene* or *Astraea* species; body whorl below shoulder and base of shell with 4 large, thick, beaded cords; smaller secondary cords between larger cords; area between suture and shoulder smooth; shell solid orange-tan in color with scattered darker tan spots on spiral cords and tips of shoulder knobs; interior of aperture white.

Etymology: Named for Mrs. Sally D. Kaicher of St. Petersburg, Florida.

Discussion: *Modulus kaicherae* is the deepest-dwelling *Modulus* species in the western Atlantic. Unlike the other known species from Florida, *M. kaicherae* inhabits coral rubble and coralline algae bottoms in 70 m depths, and is part of the unusual Palm Beach coral rubble talus fauna which includes other endemics such as *Coralliophila pacei* Petuch, *Dermomurex (Trialatella) glicksteini* Petuch, and *Conus harasewychi* Petuch. *Modulus kaicherae,* although very distinctive, is probably closest to *M. modulus* (Linnaeus), but differs in being a wider, more flattened shell, in having eight large, pointed knobs per whorl, having fewer spiral cords on the body whorl, and in being solid orange-tan in color. The smooth area between the suture and the shoulder knobs is also distinctive.

Modulus kaicherae has only been dredged from off Palm Beach, and it may be endemic to the southeastern coast of Florida. Ecologically, the Florida *Modulus* species differ greatly; *M. modulus* (Linnaeus) and *M. floridanus* (Conrad) live in and on Turtle Grass *(Thalassia)* beds and algae in very shallow, often intertidal, depths; *M. pacei* Petuch (described here) lives on algae-covered rocks in intertidal depths; *M. lindae* Petuch (described here) lives on red algae in 60 m depths; and *M. kaicherae* lives on coralline algae and coral rubble in 70 m depth. Altogether, these species constitute the richest modulid fauna in the world.

Modulus lindae new species (Plate 3, Figures 5, 6)

Material Examined: HOLOTYPE - Diameter 10 mm, height 10 mm, trawled from 60 m depth off Palm Beach Island, Palm Beach, Florida, 1981, USNM 859825; PARATYPES - 1 specimen, same locality as holotype, Petuch collection; 12 specimens, same locality as holotype, Glickstein collection; 5 specimens, 55 m depth off St. Augustine, Florida, Robert Pace collection.

Description: Shell small for genus, trochiform, with elevated spire and sloping spire whorls; shoulder sharp-angled, with single large, beaded cord around periphery; area between shoulder cord and suture with 14-20 low, beaded axial ribs; base of shell, from shoulder cord to anterior tip of columella, with 3 close-packed spiral cords; umbilicus open; columellar tooth very large in proportion to shell size; color varying from tan to pink, purple, or lilac, with most specimens being light purple; spiral cords with evenly-spaced reddish-brown dots; protoconch and early whorls purple; columella and columellar tooth bright lilac-purple; interior of aperture pale lilac or pinkish-purple; interior of aperture with numerous ribs.

Etymology: Named for my wife, Linda J. Petuch.

Discussion: *Modulus lindae* is closest to *M. modulus* (Linnaeus) (Plate 3, Figure 7) and *M. floridanus* Conrad, but differs in being a smaller shell with a consistently much higher spire and trochiform shape instead of a discoidal shape, in having a finer, more delicate sculpturing, in having a larger apertural tooth in proportion to shell size and in having the distinctive purple and lilac coloration. *Modulus lindae* also has a totally different ecology than does *M. modulus* or *M. floridanus.* The new species is found only in deeper (50-60 m) offshore water where it lives on deep water red algae and red coralline algal nodules. *Modulus lindae* ranges from near Jacksonville, Florida south to Palm Beach, Florida, at the edge of the continental shelf in red algae beds. In general shape and ornamentation, *M. lindae* somewhat resembles members of the periwinkle genus *Nodilittorina (Echinolittorina),* such as *N. tuberculata.*

Modulus pacei new species (Plate 3, Figures 1, 2)

Material Examined: HOLOTYPE - Diameter 14 mm, height 13 mm, under algae-covered rock, beachrock reef on northeastern side of Peanut Island, Palm Beach Inlet, Lake Worth, West Palm Beach, Florida, 1983, USNM 859826; PARATYPE - 1 specimen, same locality and date, USNM 859827; 6 specimens, same locality, Robert Pace collection.

Description: Shell trochiform, with elevated spire and sloping spire whorls; shoulder sharply angled, with large beaded cord around periphery; shoulder with 8-10 large axial ribs; axial ribs extend over edge of peripheral cord, giving an astraeiform appearance; shoulder area with 4 beaded spiral cords; base of shell with 4 large spiral cords; umbilicus open; columellar tooth small in proportion to shell size; columella, umbilicus, and posterior section of aperture colored brownish-purple; shell white with intermittent black spots on all spiral cords.

Etymology: Named for Mr. Robert Pace of Miami, Florida.

Discussion: *Modulus pacei* is closest to *M. modulus* (Linnaeus), but differs in being a much larger shell with a sharper shoulder, higher spire, and more trochiform shape, in having more spiral cords on both the shoulder and the base, in having the shoulder axial ribs extend beyond the periphery, giving the shell the appearance of an *Arene* or an *Astraea,* and in having jet black fleckles on a white background instead of brown patches and dots on a cream-colored background. Ecologically, the two species differ greatly; *M. modulus* is found in sea grass *(Halidule* and *Thalassia)* and weed beds in shallow water, from North Carolina to Brazil, while *M. pacei* is found on and under rocks (feeding on algal films) on rock reefs from Ft. Pierce, Florida to Palm Beach, Florida. *Modulus pacei* coexists with *Cerithium lindae* Petuch, which has the same range and habitat. This new *Modulus* is most probably an offshoot of *M. modulus* that evolved from an isolated population during the Pleistocene, in the same manner and for the same ecological and physiological reasons that *Cerithium lindae* evolved from *C. lutosum* (see Discussion under description of *Cerithium lindae). Modulus pacei* is most common at Peanut Island, Palm Beach.

Family **Muricidae**

Genus *Dermomurex* Monterosato, 1890

Subgenus *Trialatella* Berry, 1964

Dermomurex (Trialatella) glicksteini new species (Plate 3, Figures 15, 16)

Material Examined: HOLOTYPE - Length 16 mm, trawled from 150 m depth of Palm Beach Island, Palm Beach County, Florida, 1980, USNM 859836; PARATYPE - 1 specimen, same locality and depth as holotype, Glickstein collection.

Description: Shell elongated, fusiform, with elevated spire and well developed siphonal canal; whorls with 3 large, blade like varices; intervarical areas with 1 large, keeled axial nob; whorls ornamented with numerous raised spiral threads; aperture large, oval in shape; inner edge of lip with 6 large red denticles; shell color mottled khaki-green and tan, overlaid with darker tannish-brown banding along the suture and on siphonal canal; intritacalx thick, chalky, light tan in color; interior of aperture light tan.

Etymology: Named for Dr. Marvin Glickstein of North Palm Beach, Florida, who dredged the holotype.

Discussion: *Dermomurex glicksteini* is the fourth-known *Trialatella* to be described from the western Atlantic, and the first *Trialatella* species to be collected from the mainland United States. The other species in the subgenus include *D. (Trialatella) abyssicola* (Crosse) from deep water off Guadeloupe and the lesser Antilles, *D. (Trialatella) kaicherae* Petuch (described in this book) from Venezuela, and *D. (Trialatella) oxum* Petuch, 1979, from the Abrolhos Islands of Brazil. *Dermomurex (Trialatella) glicksteini* differs from all three of these species in being much more slender and elongated, by having a much higher, more protracted spire, by having the khaki-green shell color, and by having the six large red teeth on the inside of the lip in the aperture. This new *Trialatella* species has only been collected from deep water, on a coralline algae and coral rubble bottom, off Palm Beach, Florida.

Genus *Murexiella* Clench and Farfante, 1945

Murexiella taylorae new species (Plate 10, Figures 8,9)

Material Examined: HOLOTYPE - Length 16 mm, trawled from 200 m depth off Cedar Key, Florida, 1973, USNM 859844; PARATYPES - 1 specimen, same depth and locality as holotype, Kevan Sunderland collection.

Description: Shell small for genus, with oval body, elevated spire, and long siphonal canal; 7 varices per whorl; varices with 6 large, strongly-recurved spines; intervarical areas with 6 large fimbriated cords; siphonal canal with 2 large, recurved spines; last spines of siphonal canal greatly recurved, almost touching varical spines; shoulder sharply angled; spire whorls tabulate; shell color rosy-tan with dark rose spots in pits at the base of each varix; siphonal canal pinkish-rose; interior of aperture pinkish-rose.

Etymology: Named for Linda Taylor, Ft. Lauderdale, Florida.

Discussion: *Murexiella taylorae* has been recognized as a new species for years now, but was never formally described. FAIR (1976, plate 20, figures 274, 274a) illustrates a specimen of this new species from 100 fathoms off Cape San Blas, Florida and notes that it is un-named. *Murexiella taylorae* is closest to *M. macgintyi* (M. Smith), but differs in being a smaller shell with a much sharper-angled shoulder, in having tabulate, more protracted spire whorls, in having only two spines per varix on the siphonal canal, and in having the rosy color and dark rose spots behind the varices. In having a wide, sharp-angled shoulder, *M. taylorae* is similar to the angular *M. graceae* (McGinty) from the middle Pleistocene Bermont Formation of Florida. It differs from that fossil species in being smaller, more globose in shape, and having a proportionally higher spire.

As far as is known, *Murexiella taylorae* is only found off northwestern Florida, in deep water.

Genus *Muricopsis* Bucquoy, Dautzenberg, and Dollfuss, 1892

Muricopsis richardbinghami new species (Plate 5, Figures 12, 13)

Material Examined: HOLOTYPE - Length 17 mm, width 9 mm, collected by SCUBA diving, under coral rubble in 100 ft. of water off Palm Beach Island, Palm Beach, Florida, 1978, USNM 859847.

Description: Shell fusiform, elongated, with elevated spire; shell with 8 varices per whorl; body whorl ornamented with 5 large, evenly-spaced, raised cords; siphonal canal with 4 cords; spire whorls with 3 spiral cords; shoulder tabulate, sharp-angled; varices thin, winglike, with 5 large, bladelike serrations; serrations correspond to cords in intervarical regions; posteriormost serration largest, spinelike, pointed posteriorly; aperture small, oval; outer lip thickened, with 5 large denticles along inner edge; shell color tan, with 2 thin reddish-brown bands, 1 around subsutural region and 1 around break between body whorl and siphonal canal; spire whorls with reddish-brown subsutural band; interior of aperture pale tan.

Etymology: Name for Mr. Richard Bingham, who collected the holotype off Palm Beach.

Discussion: *Muricopsis richardbinghami* is unlike any other Caribbean muricopsine. This new species is closest to *M. oxossi* Petuch (Plate 15, Figures 15, 16) from the Abrolhos Archipelago off southern Brazil, but differs in being more elongated, in lacking the scabrous fimbriations between the varices, and in having a darker color with two dark bands. *Muricopsis blainvillei* (Payraudeau) from the Mediterranean Sea is also similar, but is a larger, much more slender species.

Muricopsis richardbinghami is a rare species, and appears to inhabit the deep water coral rubble zones off Palm Beach County. This poorly-explored area contains other endemic species such as *Conus harasewychi* Petuch and *Dermomurex (Trialatella) glicksteini* Petuch, all of which probably occur together, buried deeply within the coral rubble.

Family **Coralliophilidae**

Genus *Coralliophila* H. and A. Adams, 1853

Coralliophila pacei new species (Plate 3, Figures 11, 12)

Material Examined: HOLOTYPE - Length 8 mm, width 6 mm, deeply buried under coral rubble in 60 ft. depth, by SCUBA diver, off Pompano outfall, Pompano Beach, Florida, 1981, USNM 859850; PARATYPE - 1 specimen, same locality as holotype, Robert Pace collection.

Description: Shell small for genus, with ovate, inflated body whorl and elevated, scalariform spire; body whorl ornamented with 2 large, fimbriated spiral cords, one at shoulder and one below (anterior to) mid-body; 3 small, squamose spiral cords between suture and large shoulder cord; 2 small, squamose spiral cords between 2 large cords around anterior end; 9 low varices per whorl; large scale produced at intersection of varix and spiral cord, giving shell fimbriated appearance; shell uniformly cream-white in color.

Etymology: Named for Mr. Robert Pace of Miami Florida, who collected the holotype while SCUBA diving.

Discussion: *Coralliophila paceii* is the smallest-known Caribbean *Coralliophila,* and the only one to have a bicarinated appearance, with two large frilly cords around the body whorl. In the western Atlantic, only *Coralliophila caribaea* Abbott is close to *C. pacei.* The new Florida species differs from *C. caribaea* in being much smaller with a narrower, more protracted and scalariform spire, in having the two large, prominent cords on the body whorl, and in lacking the violet color in the aperture. *Coralliophila pacei* was found buried deeply in large coral rubble where it was living in association with colonial zoantherians. Also living interstitially deep within the coral rubble were *Mitra pallida* Usticke and *Conus granulatus* Linnaeus.

Genus *Quoyula* Iredale, 1912

Quoyula kalafuti new species (Plate 4, Figures 1, 2)

Material Examined: HOLOTYPE - Length 14 mm, width 7 mm, attached to base of *Porites* coral, 2 m depth on reef off Harry Harris State Park, southeastern end of Key Largo, Florida Keys, 1978, USNM 859851; PARATYPE - 1 specimen, Key Largo, Robert Pace collection.

Description: Shell somewhat fusiform in shape, with inflated body whorl and flaring lip; spire elevated; body whorl and spire whorls ornamented with numerous raised spiral threads, which give shell rough appearance; aperture very wide and flaring; columella with large, flaring parietal shield; parietal shield detached from shell along edges; shoulder rounded; shell color pale lilac; interior of aperture lilac-purple; columella white.

Etymology: Named for Mr. Theodore Kalafut, of Key West, Florida, who collected the holotype.

Discussion: *Quoyula kalafuti* is the first member of this Indo-Pacific and Panamic genus to be found in the western Atlantic. Like its closely-related Pacific relative, *Q. madreporarum* (Sowerby), *Q. kalafuti* also appears to be variable in outline, with the shells taking on and conforming to the shape of the coral colony upon which they are living. This variability in shell shape is quite striking, as the paratype of *Q. kalafuti* has a much lower spire than the holotype, and closely resembles one of the variants of *Q. madreporarum, Q. madreporarum* form *monodata* (Quoy and Gaimard). The new Caribbean analogue species differs from the Pacific *Q. madreporarum* in being a more slender shell with much coarser sculpturing, and in lacking the dark purple coloring on the columella. To date, *Quoyula kalafuti* is only known from *Porites* coral beds in the lagoonal areas behind the main reef tract off the Florida Keys. Here, it is found attached to the bases of the coral colony, where it feeds on the coral animals and the mucus that they exude.

Family **Columbellidae**
Genus *Zafrona* Iredale, 1916

Zafrona taylorae new species (Plate 10, Figures 6,7)

Material Examined: HOLOTYPE - Length 11 mm, width 5 mm, on *Thalassia* (Turtle Grass) blade, in grass beds in Rabbit Key Basin, off Rabbit Key, Florida Bay, 2 m depth, Everglades National Park, 1983, USNM 859858; PARATYPES - 6 specimens, same locality and depth as holotype, USNM 859859; 20 specimens, Rabbit Key Basin, 2 m depth, Kevan Sunderland collection; 15 specimens, Rabbit Key Basin, Robert Pace collection.

Description: Shell fusiform, elongated, with elevated spire; shoulder and spire whorls rounded; shell sculptured with numerous microscopic spiral threads, giving shell shiny, silky texture; 3 or 4 spiral threads near suture larger than others; body whorl without axial sculpture; columellar callus raised, with edge detached from shell; columella with 1 large, prominent plication; inner edge of lip bordered with 8-12 small teeth; shell translucent pale greenish-tan overlaid with dense, light brown net pattern; holes in mesh of net pattern oval in shape, regularly spaced; some specimens with thin, clear band around mid-body; protoconch and early whorls dark purple-brown; columella and edge of lip white; net pattern showing through in interior of aperture.

Etymology: Named for Linda Taylor of Ft. Lauderdale, Florida.

Discussion: *Zafrona taylorae* is closest to the widespread Caribbean *Z. pulchella* (Blainville), but differs in being a smoother shell without prominent axial sculpturing, in lacking the large brown flammules and mottlings of *Z. pulchella* and having, instead, a close-meshed net pattern, in having a pale greenish-tan base color instead of white, in having a dark purple-brown protoconch and early whorls, and in having a larger plication on the columella. The ecologies of the two species also differ greatly; *Z. pulchella* lives under rocks and rubble in shallow water and on coral reefs, while *Z. taylorae* lives exclusively on Turtle Grass blades, where its greenish-tan color helps it to blend in with the rich epifaunal and epifloral community. *Zafrona taylorae* appears to be restricted to the Florida Bay section of the Everglades National Park, where it lives along with the other Florida Bay endemics, *Argopecten irradians taylorae* Petuch and *Haminoea taylorae* Petuch.

Family **Olividae**
Genus *Oliva* Bruguière, 1798
Subgenus *Strephona* Mörch, 1852

Oliva bifasciata sunderlandi new subspecies (Plate 3, Figures 13, 14)

Material Examined: HOLOTYPE - Length 22 mm, trawled from 150 m depth, west of Cedar Key,

Florida, 1977, USNM 859904; PARATYPES - 3 specimens lengths 21-23 mm, Petuch collection, same locality as holotype; 5 specimens, same locality as holotype, Sunderland collection; 1 specimen, same locality as holotype, Sargent collection; 1 specimen, same locality as holotype, Leonard Hill collection.

Description: Shell consistently smaller than nominate subspecies, stocky, with rounder sides; spire elevated; shell color pale yellow orange overlaid with dense network of tiny, orange-tan triangles; body whorl encircled with 2 wide bands of brown dots and zig-zag flammules; spire enamel pale orange-tan; protoconch yellow; interior of aperture cream-yellow; edge of suture marked with numerous fine, brown axial hairlines; columella plications white.

Etymology: Named for Mr. Kevan Sunderland, Ft. Lauderdale, Florida, who kindly donated the type material.

Discussion: *Oliva bifasciata sunderlandi* is the fourth-known subspecies of an interesting complex that is centered around *Oliva bifasciata* Küster. Members of this group include *O. bifasciata bifasciata* from the Bahamas, Cuba, and the West Indies, *O. bifasciata bollingi* Clench, which is found along both coasts of Florida and the Florida Keys, and northward to offshore of North Carolina, and *O. bifasciata jenseni* Petuch and Sargent, which is endemic to Bermuda. All of these subspecies are found in shallow water, from intertidal depths to just offshore (PETUCH and SARGENT, 1986). *Oliva bifasciata sunderlandi*, on the other hand, lives in deep water along the northwestern edge of the Floridian continental shelf. It has been trawled along with the other deep water species such as *Fasciolaria bullisi* Lyons, *Busycon lyonsi* Petuch, and *Heilprinia timessus* (Dall), and is a characteristic component of that outer continental shelf ecosystem.

Oliva bifasciata sunderlandi differs from the other three subspecies of *O. bifasciata* in being a consistently much smaller shell (averaging 22 mm as opposed to 40 mm for the others) with a more rounded, stocky appearance, in being proportionally shorter and stumpier, and in being pale orange in color. This new subspecies is restricted to deep water along the outer edge of the northwestern continental shelf area of Florida, in the Gulf of Mexico.

Family **Conidae**

Genus *Conus* Linnaeus, 1758

Conus binghamae new species (Plate 5, Figures 1, 2)

Material Examined: HOLOTYPE - Length 18 mm, width 9 mm, trawled from 200 ft. depth off Dania, Broward County, Florida, 1982, USNM 859876; PARATYPES - length 19 mm, trawled from 200 ft. depth off Palm Beach, Florida, Glickstein collection; 6 specimens, same depth and locality as holotype, Bingham collection; 1 specimen, same depth and locality as holotype, Kremer collection.

Description: Shell small for genus, thin and delicate; spire low, with early whorls protracted; body whorl shiny, sculptured with fine spiral cords; spiral cords become stronger and larger around anterior one-third of body whorl; shoulder sharply-angled; aperture narrow; shell color pattern comprising darkly-colored anterior one-third and mid-body band of large, square-shaped flammules; unpatterned posterior one-half of body whorl with rows of tiny dots; color varying from red, orange, apricot-yellow, to pink and bluish-purple (holotype reddish-orange with darker red-orange mid-body band and anterior tip); spire whorls with numerous crescent-shaped flammules; aperture of holotype dark red-orange; protoconch and early whorls bright yellow on all specimens, regardless of body whorl color.

Etymology: Named for Mrs. Patricia Bingham of Palm Beach, Florida, who recognized the species as new and kindly donated the holotype.

Discussion: *Conus binghamae* is closest to the sympatric *C. patae* Abbott, in both shape and size, but differs in being a smoother shell by lacking axial plications and heavy beading, and by having a different color pattern, consistently composed of a checkered mid-body band and darker anterior tip. The base color of *C. binghamae* is also more variable than that of *C. patae,* which is consistently in shades of pink or lavenders with dark orange scattered flammules. *Conus patae* also lacks rows of tiny dots seen on *C. binghamae.* While the two species occur together along eastern Florida, *C. patae* is more wide-ranging, being found in Jamaica, Honduras, and the Virgin Islands, while *C. binghamae* is known only from southeastern Florida.

Conus floridanus patglicksteinae new subspecies (Plate 5, Figures 3, 4)

Material Examined: HOLOTYPE - Length 25 mm, width 13 mm, trawled from 400 ft. depth off Palm Beach Island, Palm Beach County, Florida, 1982, USNM 859980; PARATYPE - 1 specimen Glickstein collection.

Description: Shell solid, elongate, tapered, with low spire; body whorl smooth, with only a few faint spiral sulci around anterior tip; shoulder sharp-angled; aperture long and narrow; color bright pinkish-orange with band of scattered white patches around mid-body and few scattered white patches along shoulder area; orange body color overlaid with 12 rows of large orange-brown dots and dashes; spire whorls white with large, bright orange, interconnected flammules; interior of aperture pale pinkish-white.

Etymology: Named for Mrs. Pat Glickstein of North Palm Beach, Florida.

Discussion: This new deep water subspecies of the common shallow water *Conus floridanus* Gabb (and its highly colored variety *floridensis* Sowerby) differs from the nominate species in having a low, almost flattened spire, in being almost solid orange-pink in color, and in having large, radiating, interconnected orange spire flammules. In many ways, particularly the low, flattened spire and rows of large dots, *C. floridanus patglicksteinae* resembles certain orange varieties of *C. regularis* Sowerby from the Gulf of California. The new subspecies differs from *C. floridanus burryae* Clench from the Florida Keys (a full species?) by having a much lower, flattened spire and by having a bright pinkish-orange base color. *Conus floridanus patglicksteinae* lives together with the other southeastern Florida deep water cones, *C. flamingo* Petuch, *C. binghamae* Petuch, *C. glicksteini* Petuch, and *C. amphiurgus* Dall (= *C. juliae* Clench).

Conus glicksteini new species (Plate 5, Figures 5, 6)

Material Examined: HOLOTYPE - Length 21 mm, width 11 mm dredged from 400 ft. depth off Palm Beach Island, Palm Beach County, Florida, 1982, USNM 859881; PARATYPES - 4 specimens, lengths 12-22 mm, same depth and locality as holotype, Glickstein collection.

Description: Shell small for genus, thin, delicate, with low spire; body whorl smooth and shiny, with on few weak spiral threads around anterior tip; color varying from salmon-pink to pinkish-lavender, with evenly spaced pale tan lines or rows of dots around body whorl (holotype salmon-pink with only few rows of pale tan dots around mid-body); all specimens with paler band around mid-body and around shoulder; spire whorls with numerous pale orange, thin, crescent-shaped flammules; interior of aperture pink; protoconch and early whorls bright pink in all specimens, regardless of body whorl color; aperture narrow, shoulder slightly rounded.

Etymology: Named for Dr. Marvin Glickstein of North Palm Beach, Florida, who collected the holotype and paratypes during his amateur dredging operations in deep water off Palm Beach.

Discussion: *Conus glicksteini* is a rather undistinguished little species, but with a subtle beauty all of its own. The holotype very closely resembles a small specimen of the Japanese species *C.*

kashiwajimensis Shikama, particularly in having a slightly rounded shoulder and having broad bands of salmon-pink on a lighter salmon background. The lavender-colored paratype (22 mm) somewhat resembles *C. floridulus* Adams and Reeve from deep water in the southwestern Pacific which, in turn, is also related to *C. kashiwajimensis.* In the western Atlantic, *C. glicksteini* is closest to *C. daucus* Hwass, but differs in being a much smaller shell with a more rounded shoulder and higher spire. *Conus glicksteini* is also close to *C. attenuatus* Reeve, but differs in being a stockier, less elongated shell and by having a color pattern of wide salmon-pink or lavender bands overlaid with rows of tan lines or rows of dots. *Conus daucus, C. attenuatus,* and the related *C. flavescens* Sowerby, are all relatively shallow water species, being collected in 10 to 60 ft. depths along southeastern Florida. *Conus glicksteini,* on the other hand, is a deep water species, being trawled from 400 ft. depths, and lives with the other deep water eastern Floridian cones, *C. flamingo* Petuch, *C. amphiurgus* Dall (= *C. juliae* Clench), and *C. binghamae* Petuch.

Conus harasewychi new species (Plate 5, Figures 7, 8)

Material Examined: HOLOTYPE - Length 26 mm, width 14 mm, collected by SCUBA diver on coral rubble bottom, 30 m depth, north of Palm Beach Inlet, Palm Beach, Florida, 1974, USNM 859882; PARATYPE - length 24 mm, off Palm Beach, Florida, collection of Charles Key.

Description: Shell stocky, solid, broad across shoulder; shoulder somewhat rounded, distinctly coronated with large raised knobs; body whorl with numerous fine, beaded, spiral cords; spiral cords become stronger on anterior half of shell; body whorl uniformly dark mustard-yellow with thin, amorphous white band around mid-body; shoulder coronations and spire whorls white in color; anterior tip of shell and siphonal region dark brown; interior of aperture pale lavender.

Etymology: Named for Dr. M. G. Harasewych of the Division of Mollusks, NMNH, Smithsonian Institution.

Discussion: *Conus harasewychi* is closest to *C. cardinalis* Hwass, in shape, size, and sculpturing, but differs by lacking the well-defined, dark-flammuled and spotted mid-body band, and by lacking the dark, hairline flammules on the spire whorls. Within the *C. cardinalis* species complex, the pure white, unicolored spire of *C. harasewychi* is unique. From the closely-related Bahamian endemic, *C. jucundus* Sowerby (= *C. abbotti* Clench), *C. harasewychi* differs in being a much smaller, heavier, and stockier shell, and by having a dark-colored anterior tip. In this last characteristic, the new species is similar to the Roatan Island endemic *C. kulkulcan* Petuch. *Conus harasewychi* is the first member of the *C. cardinalis-magellanicus* species complex to be found on the mainland of the United States.

Order **Cephalaspidea**
Superfamily **Bullacea**
Family **Haminoeidae**
Genus *Haminoea* Turton and Kingston, 1830

Haminoea taylorae new species (Plate 4, Figures 12, 13)

Material Examined: HOLOTYPE - Length 16 mm, width 12 mm, on mud flat at edge of mangrove forest, low tide, Joe Kemp Key, Florida Bay, Everglades National Park, 1983, USNM 859897; PARATYPES - 4 specimens, same locality and date as holotype, USNM 859898; 10 specimens, Kevan Sunderland collection.

Description: Shell fragile, glassy, translucent, inflated; shape oval-cylindrical; outer lip only slightly curved; shell ornamented with numerous very fine spiral, thread-like grooves; anterior half of aperture 3 times as wide as posterior half; columella greatly curved at anterior end; spire depressed, with apical pit; shell color pale pink; juvenile specimens darker pinkish-red.

Etymology: Named for Ms. Linda Taylor, who recognized the species as new.

Discussion: *Haminoea taylorae* is similar to the widespread Caribbean *H. elegans* (Gray), but differs in being a larger, more cylindrically-shaped shell, by having a much narrower posterior half of the aperture in proportion to anterior half, by having a much more recurved columella, and in having a pink shell instead of a yellowish-white one. *Haminoea taylorae* is also similar to another widespread Caribbean species, *H. antillarum* (Orbigny), but differs in being a more slender shell with a straighter outer lip, in having fine spiral grooves instead of being smooth, and in being pink instead of greenish-yellow. *Haminoea antillarum,* the tiny, cylindrical *H. succinea* (Conrad), and *H. taylorae* occur together in Florida Bay, often being trawled together from *Thalassia* (Turtle Grass) beds and *Halidule* beds near mud flats adjacent to mangrove islands. *Haminoea taylorae* is apparently endemic to the Florida Bay section of the Everglades National Park.

Class **Bivalvia**
Order **Pterioida**
Suborder **Pteriina**
Superfamily **Pectinacea**
Family **Pectinidae**
Genus *Argopecten* Monterosato, 1889

Argopecten irradians taylorae new subspecies (Plate 5, Figures 10, 11)

Material Examined: HOLOTYPE - Length 45 mm, width 46 mm, collected from try net trawling, from 10 ft. depth, on solid Turtle Grass *(Thalassia testudinum)* bottom, in Rabbit Key Basin, Florida Bay, Florida Keys, 1983, USNM 859901; PARATYPES - 8 specimens, lengths 43-28 mm, same depth, locality, and date as holotype, USNM 859902; 7 specimens, same locality, depth, and date as holotype, Petuch collection; 5 specimens, same locality as holotype, Sunderland collection.

Description: Shell thin, fragile, moderately inflated, with lower valve slightly more convex than upper valve; ribs square in cross-section in small juveniles, rapidly becoming wider and flatter as shell matures; mature specimens with very flattened, close-packed ribs; rib count ranging from 23-25 (holotype with 23); color of upper valve composed of mottled streaks and patches of dark brown, pink, yellow, and white on khaki green and gray background; some specimens with 3-5 white or yellow rays; bottom valve yellow or yellow-white, copiously covered with brown and white mottlings and flammules; in some specimens, entire bottom valve is darkly colored with brown mottlings; auricles of some specimens marked with brown and white zebra-like flammules.

Etymology: Named for Ms. Linda Taylor of Ft. Lauderdale, Florida.

Discussion: Prior to the discovery of this isolated subspecies, there were no known populations of *Argopecten irradians* found in the Florida Keys area and southern tip of Florida (ABBOTT, 1974, pp. 447-448). There are now four known living subspecies of *Argopecten irradians: A. irradians irradians* (Lamarck), which ranges from Cape Cod to New Jersey; *A. irradians concentricus* (Say), which has a split range with an eastern population ranging from New Jersey to Georgia, and western population in the Gulf of Mexico, ranging from Tampa, Florida to Louisiana; *A. irradians taylorae* Petuch, which is restricted to Florida Bay (Everglades National Park) and the western sides of the middle and upper Florida Keys; and *A. irradians amplicostatus* Dall, which ranges from central Texas south to the eastern coast of Central America. An extinct subspecies, *A. irradians sablensis* Clarke, was described from Holocene fossils from Nova Scotia.

Argopecten irradians taylorae is restricted to the dense, well developed Turtle Grass *(Thalassia testudinum)* beds that virtually carpet most of the shallow Florida Bay west of the Florida Keys. Here, it is an abundant animal, often occurring in large shoals. *Argopecten irradians taylorae* differs from *A. irradians concentricus,* its closest relative, in being a much smaller, more fragile shell, by having more ribs (23-25 as opposed to 19-21), by having ribs that are flat in cross-section as opposed to square, by having wider, more flaring ribs, and by being a much more colorful shell, with rays and mottlings of pink, brown, yellow, white, khaki, and gray. In being such a colorful, attractive small scallop, most local collectors had misidentified this subspecies as *Argopecten nucleus* (Born). The bottom valve of *Argopecten irradians taylorae* is generally yellow instead of white, as in the other subspecies, and tends to have more brown mottlings than *A. irradians concentricus.* Fossil specimens of *A. irradians taylorae* are common in Pleistocene formations in the Everglades region of southern Florida.

PLATE 1. NEW GASTROPODS FROM THE TAMPAN SECONDARY RELICT POCKET

Figures 1, 2: *Turritella (Torcula) lindae* Petuch, n. sp., dorsal and ventral views of holotype, 24 mm.

Figures 3, 4: *Cyphoma lindae* Petuch, n. sp., dorsal and ventral views of holotype, 18 mm.

Figures 5, 6: *Favartia lindae* Petuch, n. sp., dorsal and ventral views of holotype, 11 mm.

Figures 7, 8: *Murexiella kalafuti* Petuch, n. sp., dorsal and ventral views of holotype, 16 mm.

Figures 9, 10: *Chicoreus rachelcarsonae* Petuch, n. sp., dorsal and ventral views of holotype, 43 mm.

Figures 11, 12: *Busycon (Busycoarctum) lyonsi* Petuch, n. sp., dorsal and ventral views of holotype, 132 mm.

Figure 13: *Busycon (Busycoarctum) lyonsi* Petuch, n. sp., ventral view of paratype, 135 mm.

1
2
3
4
5
6
7
8
9
10
11
12
13

PLATE 2. NEW GASTROPODS FROM THE TAMPAN SECONDARY RELICT POCKET

Figures 1, 2: *Heilprinia lindae* Petuch, n.sp., dorsal and ventral views of holotype, 143 mm.

Figures 3, 4: *Muricantharus harasewychi* Petuch, n.sp., dorsal and ventral views of holotype, 25 mm.

Figures 5, 6: *Cancellaria richardpetiti* Petuch, n.sp., dorsal and ventral views of holotype, 31 mm.

Figures 7, 8: *Conus aureonimbosus* Petuch, n.sp., dorsal and ventral views of holotype, 27 mm.

Figures 9, 10: *Splendrillia sunderlandi* Petuch, n.sp., dorsal and ventral views of holotype, 40 mm.

Figures 11, 12: *Terebra (Myurella) lindae* Petuch, n.sp., dorsal and ventral views of holotype, 63 mm.

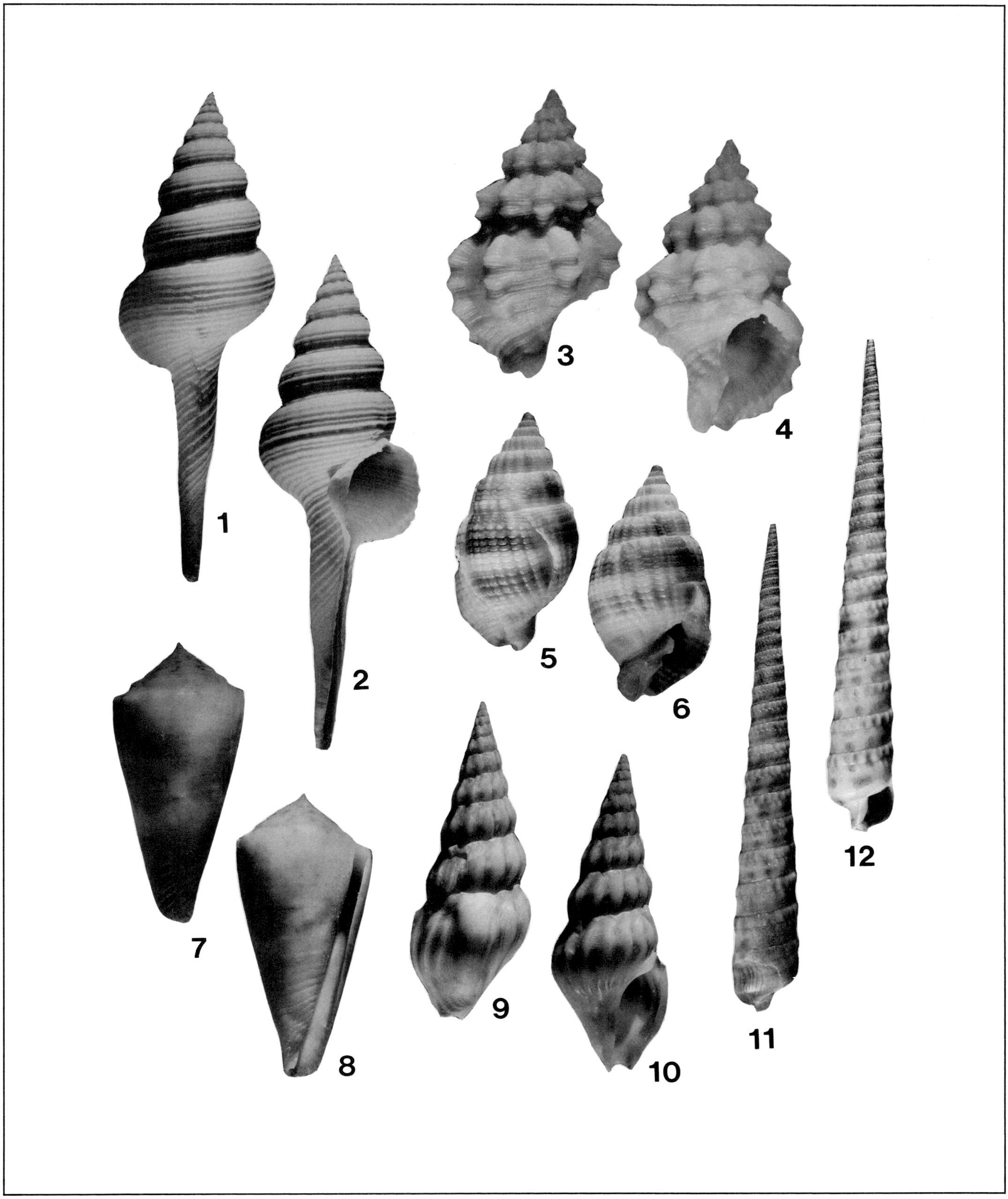
1
2
3
4
5
6
7
8
9
10
11
12

PLATE 3. NEW GASTROPODS FROM FLORIDA AND THE BAHAMAS

Figures 1, 2: *Modulus pacei* Petuch, n. sp., dorsal and ventral views of holotype, 14 mm.

Figures 3, 4: *Modulus kaicherae* Petuch, n. sp., dorsal and ventral views of holotype, 14 mm.

Figures 5, 6: *Modulus lindae* Petuch, n. sp., dorsal and ventral views of holotype, 10 mm.

Figure 7: *Modulus modulus* (Linnaeus, 1758), ventral view of 11 mm specimen, Rabbit Key, Florida Bay.

Figures 8, 9: *Cerithium lindae* Petuch, n. sp., dorsal and ventral views of holotype, 11 mm.

Figure 10: *Cerithium lindae* Petuch, n. sp., ventral view of paratype, 11 mm, Peanut Island, Palm Beach.

Figures 11, 12: *Coralliophila pacei* Petuch, n. sp., dorsal and ventral views of holotype, 8 mm.

Figures 13, 14: *Oliva bifasciata sunderlandi* Petuch, n. subsp., dorsal and ventral views of holotype, 22 mm.

Figures 15, 16: *Dermomurex (Trialatella) glicksteini* Petuch, n. sp., dorsal and ventral views of holotype, 16 mm.

Figures 17, 18: *Cerithium eburneum aliceae* Petuch, n. subsp., dorsal and ventral views of holotype, 14 mm.

Figures 19, 20: *Cerithium muscarum pacei* Petuch, n. subsp., dorsal and ventral views of holotype, 16 mm.

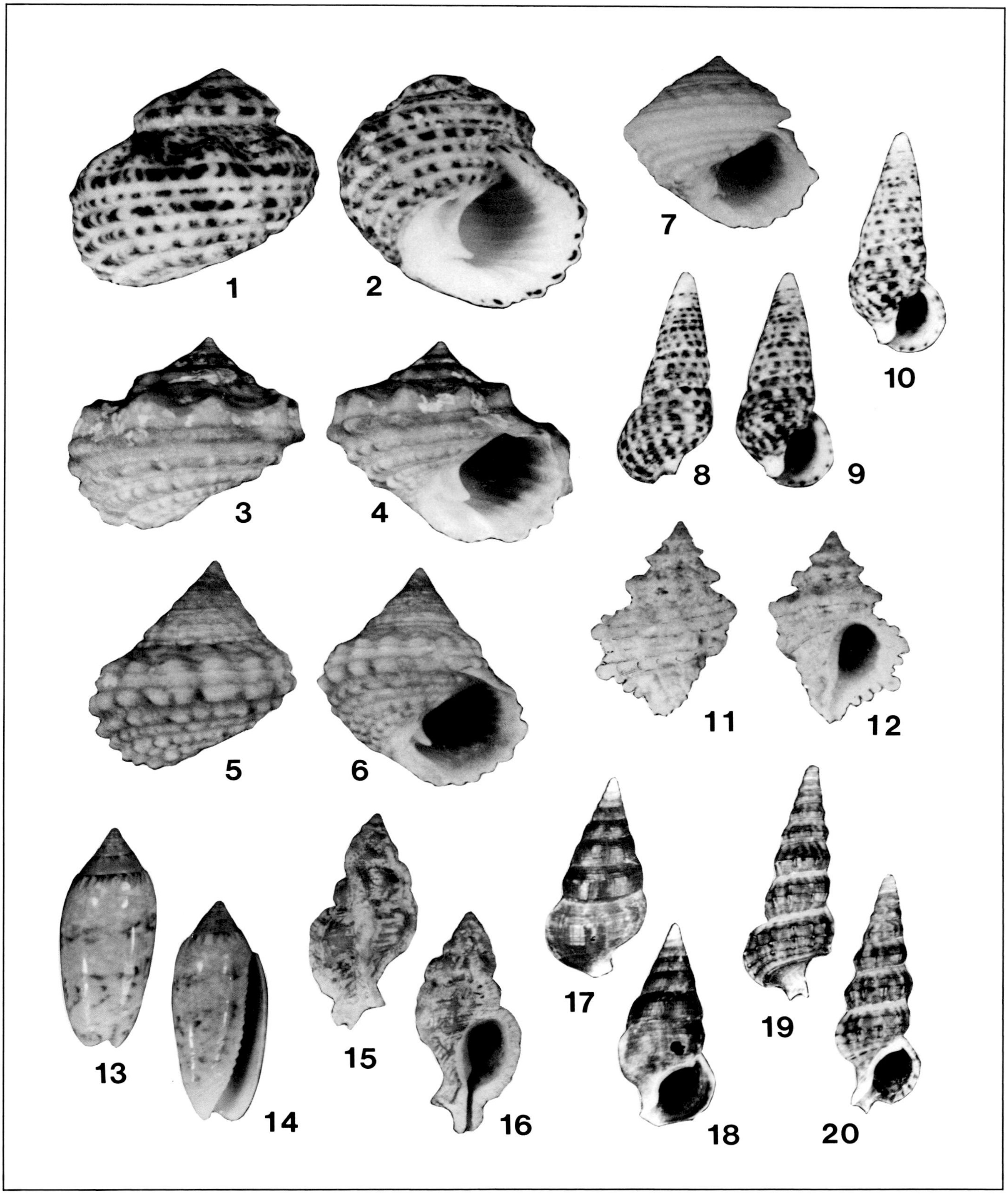
1
2
3
4
5
6
7
8
9
10
11
12
13
14
15
16
17
18
19
20

PLATE 4. NEW AND SELDOM-SEEN MOLLUSKS FROM FLORIDA AND THE SOUTHEASTERN UNITED STATES

Figures 1, 2: *Quoyula kalafuti* Petuch, n.sp., dorsal and ventral views of holotype, 14 mm.

Figure 3: Bottom valve of the scallop *Chlamys sentis,* showing attachment area (dark spot) of *Cyclotheca pacei* (Figures 6, 7).

Figures 4, 5: *Cymatium pharcidum* Dall, 1889, dorsal and ventral views of 26 mm specimen, 85 m depth off Palm Beach, Florida.

Figures 6, 7: *Cyclotheca pacei* Petuch, n.sp., dorsal and ventral views of holotype, 9 mm.

Figures 8, 9: *Coralliophila aedonis* (Watson, 1886) dorsal and ventral view of 17 mm specimen, 85 m depth off Palm Beach, Florida.

Figure 10: *Conus aureonimbosus* Petuch, n.sp., dorsal view of holotype for comparison with *Conus amphiurgus,* 27 mm.

Figure 11: *Conus amphiurgus* Dall, 1889, (= *Conus juliae* Clench, 1942), ventral view of 38 mm specimen, off St. Augustine, Florida.

Figures 12, 13: *Haminoea taylorae* Petuch, n.sp., dorsal and ventral views of holotype, 16 mm.

Figure 14: *Muricantharus harasewychi* Petuch, n.sp., paratype, 20 mm, northwestern Florida.

Figures 15, 16: *Fulgurofusus timor* Harasewych, 1983, 45 mm specimen, off Cape Fear, North Carolina.

Figure 17: *Fulgurofusus timor* Harasewych, 1983, 54 mm specimen, off North Carolina.

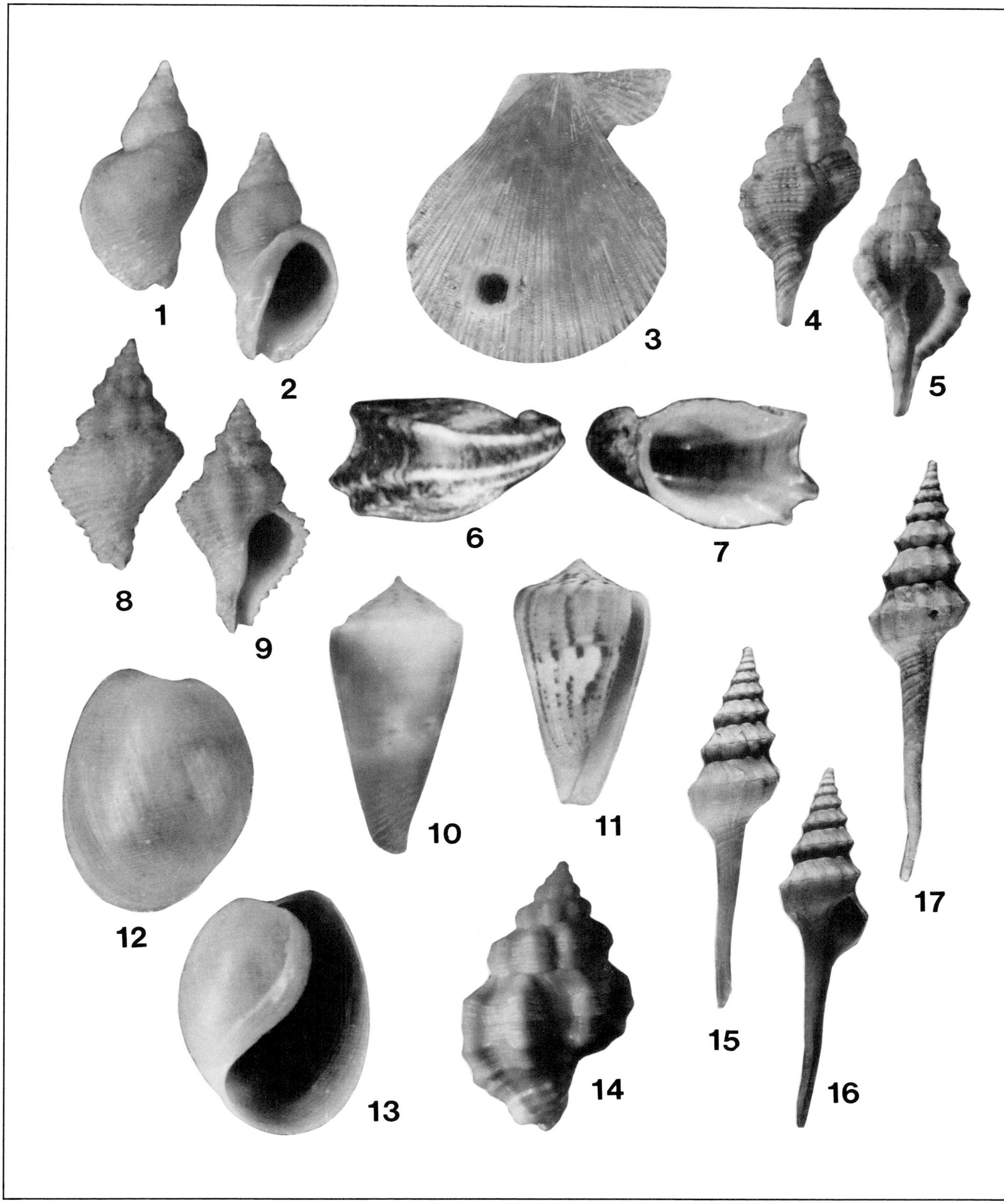
1
2
3
4
5
6
7
8
9
10
11
12
13
14
15
16
17

PLATE 5. NEW BIVALVES AND GASTROPODS FROM FLORIDA

Figures 1, 2: *Conus binghamae* Petuch, n.sp., dorsal and ventral views of holotype, 18 mm.

Figures 3, 4: *Conus floridanus patglicksteinae* Petuch, n.subsp., dorsal and ventral views of holotype, 25 mm.

Figures 5, 6: *Conus glicksteini* Petuch, n.sp., dorsal and ventral views of holotype, 21 mm.

Figures 7, 8: *Conus harasewychi* Petuch, n.sp., dorsal and ventral views of holotype, 26 mm.

Figure 9: *Argopecten irradians concentricus* (Say, 1822), upper valve, 67 mm, Emerald Isle, North Carolina.

Figures 10, 11: *Argopecten irradians taylorae* Petuch, n.subsp., dorsal and ventral views of holotype, 46 mm.

Figures 12, 13: *Muricopsis richardbinghami* Petuch, n.sp., dorsal and ventral views of holotype, 17 mm.

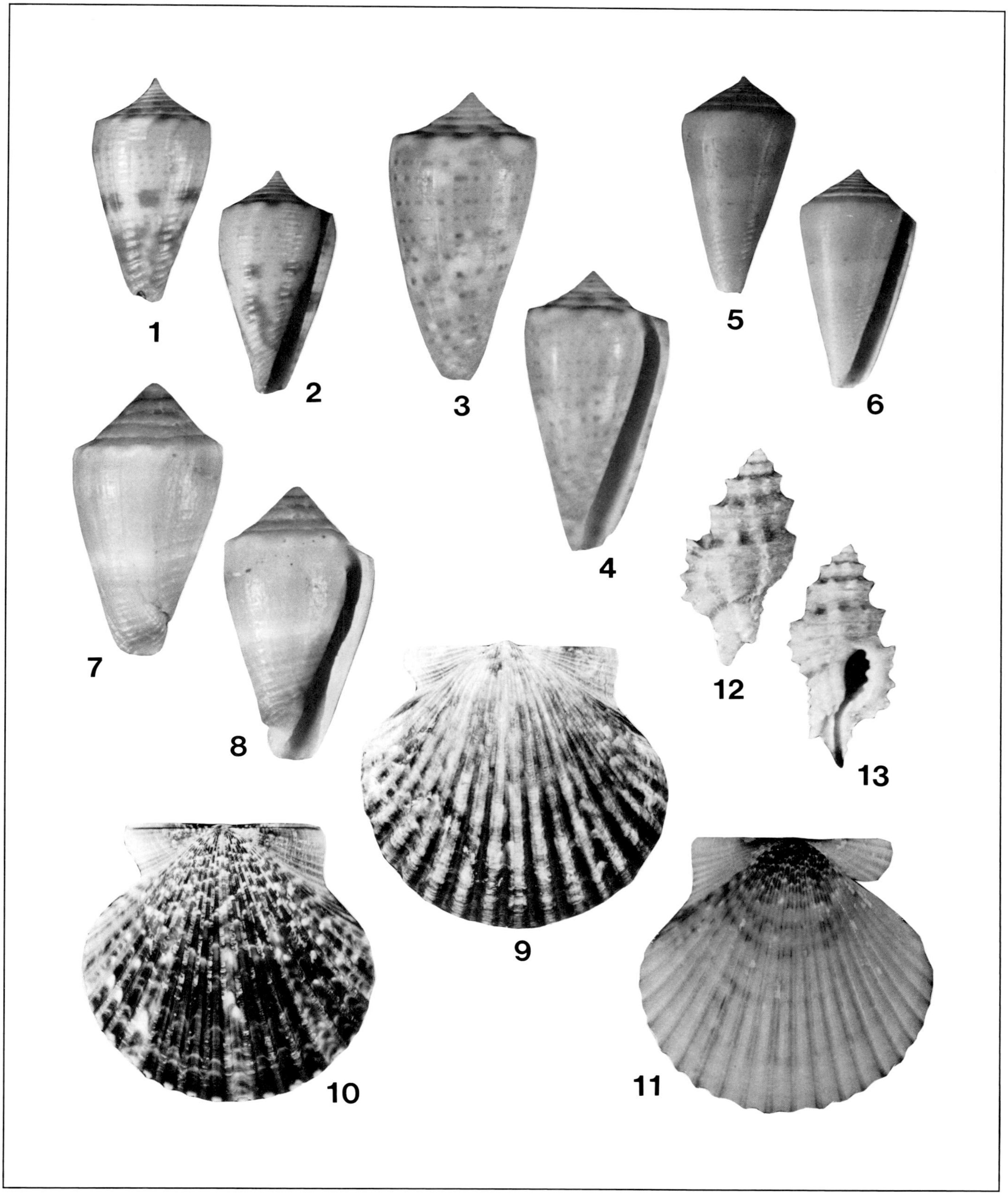
1
2
3
4
5
6
7
8
9
10
11
12
13

PLATE 6. GASTROPODS FROM THE SOUTHEASTERN UNITED STATES, BERMUDA, COZUMEL, AND FLORIDA STRAITS

Figures 1, 2: *Cypraea aubryana* Jousseaume, 1869, dorsal and ventral views of 49 mm specimen, St. Vincent Island, Lesser Antilles. Taken from Sowerby (1880). Apertural dentition apparently enhanced in this illustration.

Figures 3, 4: *Cypraea aubryana* Jousseaume, 1869 dorsal and ventral views of 45 mm specimen, off Cozumel Island, Mexico.

Figures 5, 6: *Latirus macmurrayi* Clench and Aguayo, 1941, dorsal and ventral views of holotype, 55 mm, Cuba and Straits of Florida.

Figures 7, 8: *Conus lightbourni* Petuch, 1986, dorsal and ventral views of 47 mm specimen, 180 fathoms, off Bermuda.

Figure 9: *Conus lightbourni* Petuch, 1986, dorsal view of holotype, 35 mm, 180 fathoms off Bermuda.

Figures 10, 11: *Conus flamingo* Petuch, 1980, dorsal and ventral view of holotype, 19 mm, Dania, Florida.

Figure 12: *Conus flamingo* Petuch, 1980, dorsal view of paratype, 18 mm, Dania, Florida.

Figures 13, 14: *Polystira sunderlandi* Petuch, n.sp., dorsal and ventral views of holotype, 28 mm.

Figures 15, 16: *Terebra (Strioterebrum) onslowensis* (Petuch, 1974), dorsal and ventral views of holotype, 45 mm, Onslow Bay, North Carolina.

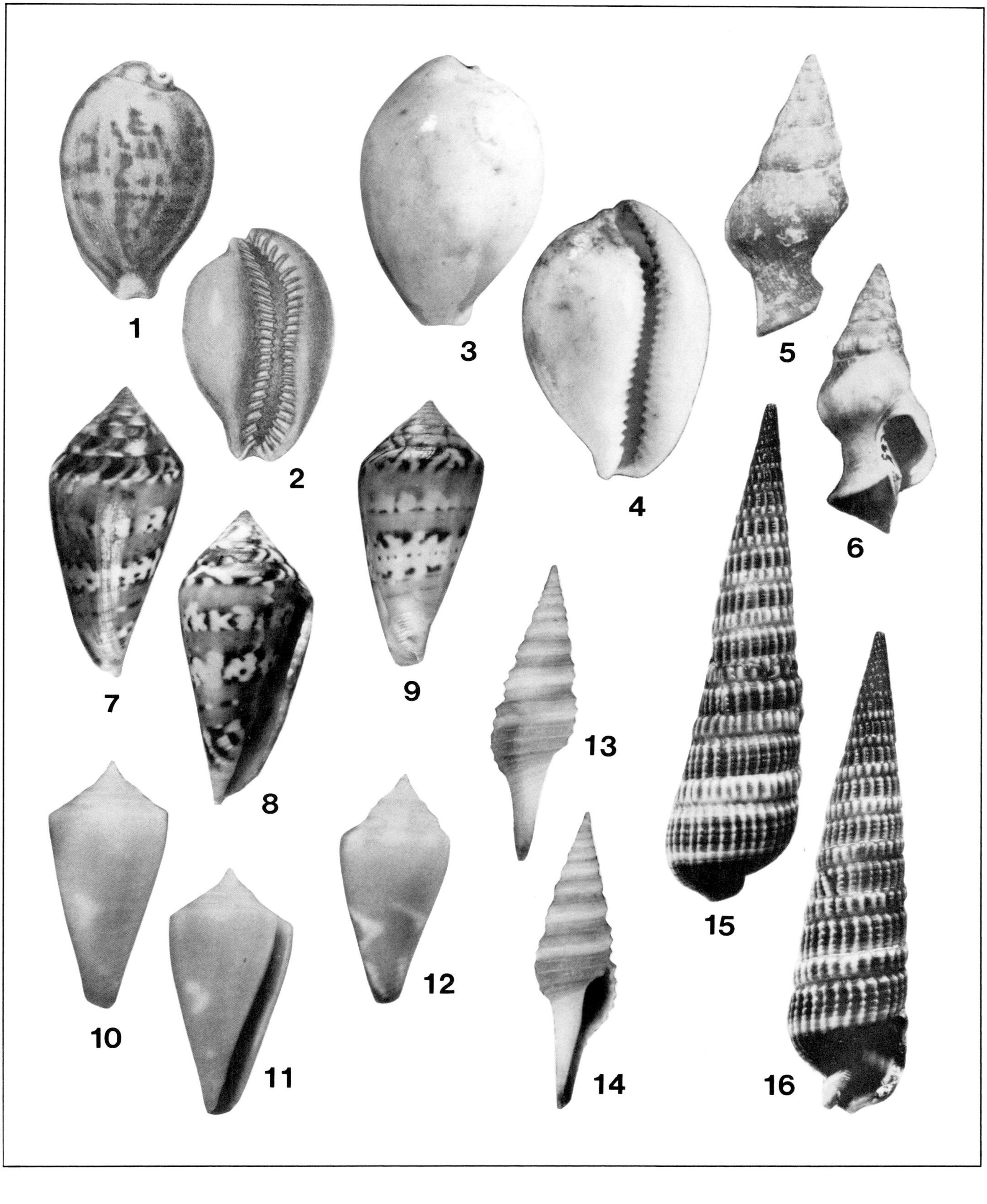
1
2
3
4
5
6
7
8
9
10
11
12
13
14
15
16

PLATE 7. DEEP WATER AND SELDOM-SEEN GASTROPODS FROM THE GULF OF MEXICO, FLORIDA, AND THE BAHAMAS

Figures 1, 2: *Fasciolaria bullisi* Lyons, 1972, holotype, northwestern Florida, av. 100 m depth.

Figure 2: *Fasciolaria branhamae* Rehder and Abbott, 1951, holotype, Campeche Bank, Mexico.

Figure 3: *Fusinus stegeri* Lyons, holotype, off western Florida.

Figure 4: *Fusinus eucosmius* (Dall, 1889), Gulf of Mexico and eastern Florida.

Figure 5: *Conus rainseae* McGinty, 1953, Campeche Bank, Mexico, 40 m depth.

Figure 6: *Conus pacei* Petuch, n.sp., Grand Bahama Island, Bahamas, 250 m depth.

Figure 7: *Conus mazei* Deshayes, 1874, St. James, Barbados, 250 m depth.

Figure 8: *Conus macgintyi* Pilsbry, 1955, Key West, Florida Keys, 200 m depth.

Figure 9: *Mitra pallida* Usticke, 1959, St. Croix, Virgin Islands and Palm Beach, Florida.

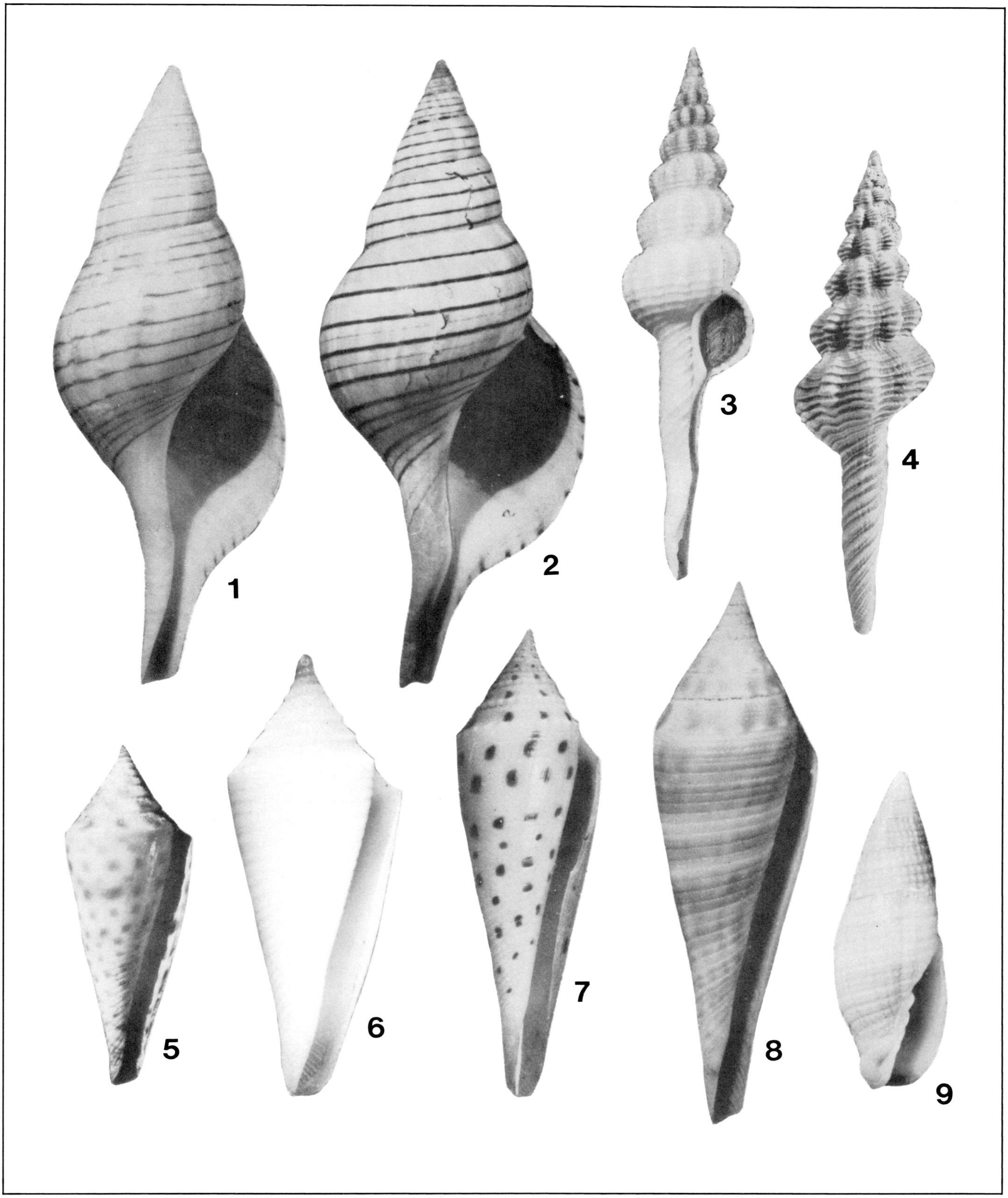
1
2
3
4
5
6
7
8
9

PLATE 8. DEEP WATER GASTROPODS FROM THE STRAITS OF FLORIDA AND THE BAHAMAS

Figures 1, 2: *Pterynotus guesti* Harasewych and Jensen, 1979, holotype, 29 mm, off Key West, Florida.

Figures 3, 4: *Emozamia richardi* (Fischer, 1882), 21 mm, Straits of Florida. (= *Latiaxis lactuca* Dall).

Figure 5: *Fulgurofusus aurora* (Bayer, 1971), holotype, 41 mm, 246 m depth off Fowey Light, Straits of Florida.

Figure 6: *Fulgurofusus merope* (Bayer, 1971), holotype, 36 mm, 549 m depth off Marquesas Keys, Straits of Florida.

Figure 7: *Fulgurofusus electra* (Bayer, 1971), holotype, 27 mm, 600 m depth off Key West, Florida.

Figure 8: *Fulgurofusus bermudezi* (Clench and Aguayo,1938), 30mm, 556 m depth off Dog Rocks, Bahamas.

Figures 9, 10: *Babelomurex fax* (Bayer, 1971), holotype, 24 mm, Straits of Florida.

Figures 11, 12: *Cordilyria cordis* (Bayer, 1971), holotype, av. 200 m depth, Hispaniola and Bahamas.

Figure 13: *Latirus varai* Bullock, 1970, holotype, av. 200 m depth, northern Cuba and Straits of Florida.

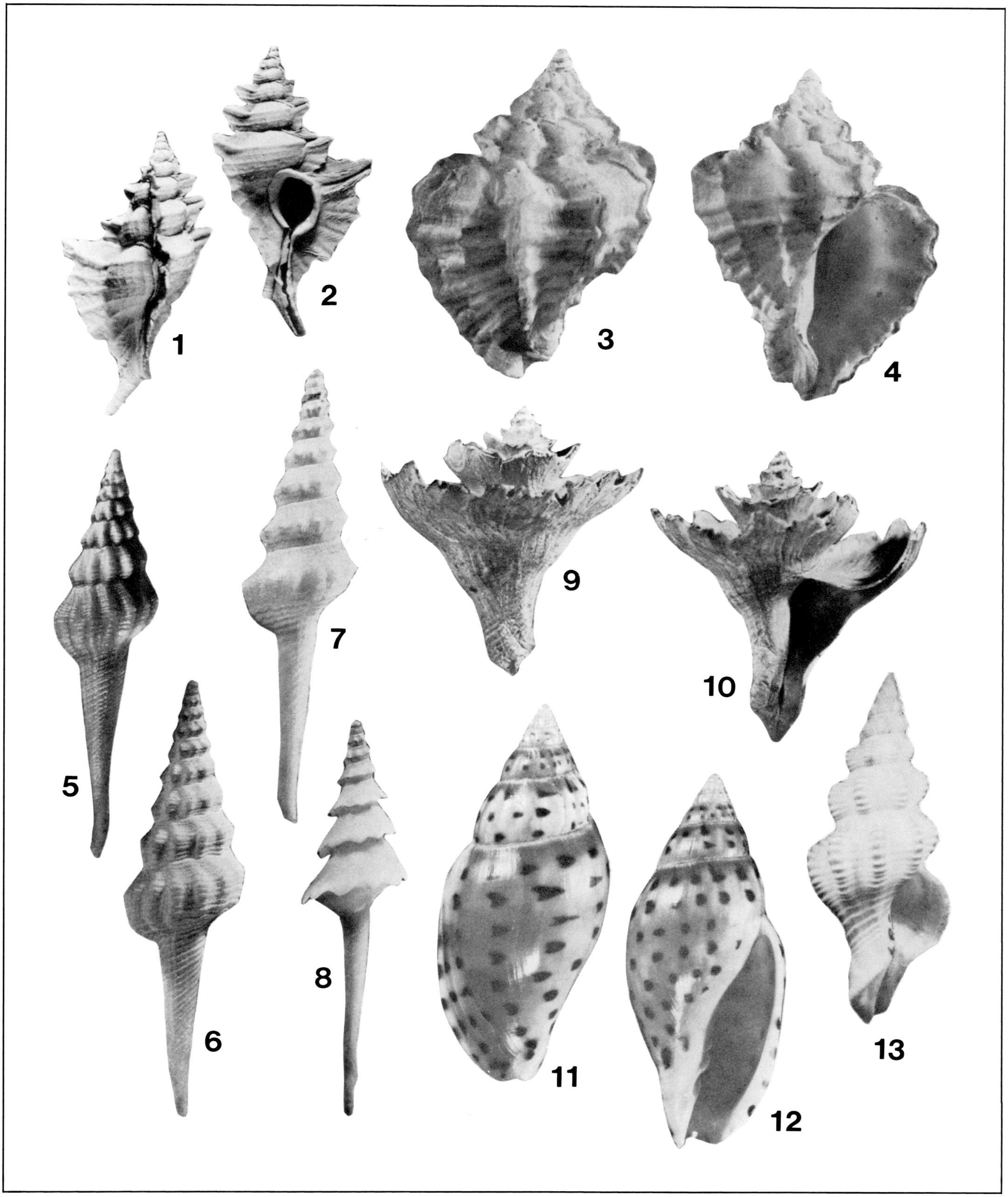
1
2
3
4
5
6
7
8
9
10
11
12
13

CHAPTER 4

New Discoveries from the Bahamas

The application of two relatively-unused collecting techniques to the sampling of the Bahamian molluscan fauna has resulted in the discovery of a number of interesting new species. These techniques include the trapping of deep water species in baited lobster pots and digging into coral rubble by SCUBA divers (previously, free diving was the only collecting method due to lack of compressors for refilling tanks on many of the out islands). As our knowledge of the Bahamian fauna grows through the efforts of divers, it is becoming more apparent that the Bahamas contain many endemic species. The species apparently evolved during the late Pleistocene when the Bahamas Platform was isolated from the rest of the Caribbean by sea level drops. Some of the Bahamian endemics that have recently been described or rediscovered include the cone shells *Conus jucundus* Sowerby (= *C. abbotti* Clench), *Conus sahlbergi* daMotta and Harland, and *Conus exquisitus* Sowerby, the costellariid *Pusia chickcharneorum* Lyons and Kaicher, and the little high intertidal cerithiid *Cerithium biminiense* Pilsbry and Brown.

The shallow water coral reef environs of the Bimini Group have recently yielded a number of interesting new species. Besides the previously-mentioned *Conus sahlbergi* (type locality South Bimini Island), two new *Terebra* species and a new, almost pitch-black, *Trivia* species have also been discovered. Along with these new species, a number of seldom-seen species appear to be relatively common in the Bimini Group. These include the volute *Lyria (Enaeta) cylleniformis* (Sowerby), the cerithiid *Fastigiella carinata* Reeve, and the miter *Mitra (Nebularia) semiferruginea* Reeve.

Collecting mollusks from baited lobster traps has added much to our knowledge of the offshore, deep water, Bahamian faunas. These traps, which have been placed in water as deep as 500 m, have brought up some of the most beautiful shells to be found in the Caribbean. In most cases, the shells are dead and occupied by hermit crabs. The crabs enter the trap to take the bait and carry along the valuable molluscan cargo with them. Some of the species collected from Bahamian waters in this manner include the coralliophilids *Babelomurex fax* (Bayer) (Plate 8, Figures 9, 10) and *Emozamia richardi* (Fischer), the turbinellid *Fulgurofusus bermudezi* (Clench and Aguayo) (Plate 8, Figure 8), the olivid *Oliva bahamasensis* Petuch and Sargent, the volutid *Volutifusus piratica* (Clench and Aguayo), and three delicate new deep water cone shells (described in this chapter).

Of particular interest, on Eleuthera Island, is a large inland salt water lake that appears to be connected to the sea by a deep subterranean stream. Originally sampled and brought to the attention of the scientific world by Mr. Robert Pace of Miami, Florida, this unusual locality contains new subspecies of *Cerithium* (described in this chapter) and a number of unusual small gastropods such as the marginellid *Cryptospira pellucida* (Pfeiffer). This salt water lake environment is discussed under the individual descriptions of the *Cerithium* species.

New Species from the Bahamas

Cerithium eburneum aliceae new subspecies
Cerithium muscarum pacei new subspecies
Trivia (Niveria) pacei new species
Chicoreus dunni new species
Muricopsis sunderlandi new species
Conus leekremeri new species
Conus lindae new species
Conus pacei new species
Terebra (Strioterebrum) biminiensis new species
Terebra (Strioterebrum) pacei new species

Family **Cerithiidae**
Genus *Cerithium*

Cerithium eburneum aliceae new subspecies (Plate 3, Figures 17, 18)

Material Examined: HOLOTYPE - Length 14 mm, width 8 mm, on algae in 1 ft. depth in enclosed salt water lake near Palmetto, Eleuthera Island, Bahamas, 1980, USNM 859818; PARATYPE - length 12 mm, same locality as holotype USNM 859819; 15 specimens, same locality as holotype Kevan Sunderland collection; 20 specimens, same locality as holotype, Robert Pace collection.

Description: Shell thin and fragile, with elongated spire but inflated body whorl; siphonal canal well-developed, projecting beyond edge of lip; shoulder slightly angled; spire whorls angled, producing turreted appearance; shell smooth, without varices or axial ribs, sculptured with numerous smooth, fine spiral threads, producing a silky texture; shoulder angle with one cord that is finely beaded; body whorl with 3-4 stronger threadlike cords, some of which may have very faint, obsolete beading; aperture wide, oval-shaped; outer lip flaring, thickened; columella covered with parietal shield; edges of parietal shield raised, not adherent to shell; shell color dark chocolate-brown with few, small, scattered white flammules on body whorl near lip; protoconch and early whorls white; columella, edge of lip, and tip of siphonal canal white; interior of aperture brown.

Etymology: Named for Mrs. Alice Pace of Miami, Florida, who collected the holotype on Eleuthera.

Discussion: *Cerithium eburneum aliceae* represents a genetically isolated population of the widespread Caribbean *C. eburneum* Bruguiere. This unusual subspecies, and the following new subspecies, are restricted to a large salt water lake, known locally as the Fishing Pond, on south-central Eleuthera Is. near the village of Palmetto. Mr. Pace (in litt.) has suggested the Fishing Pond (approximately 400 yards by 600 yards in size), which is over one-half mile inland from the ocean, may be connected by an underground channel to an offshore blue hole or flooded sinkhole. This idea is substantiated by the presence of coral reef fishes within the lake and by the abundant presence of the rare and seldom-seen marginellid *Cryptospira pellucida.* The Fishing Pond has probably been cut-off from the ocean for tens of thousands of years, possibly since the Wisconsinan II Glacial Stage of the late Pleistocene, and the very shallow water, intertidal organisms, such as *Cerithium,* would have been trapped within the lake. The tiny subterranean channel feeding into the lake would have maintained the oceanic water conditions, but would have allowed genetic drift to take place within the entrapped marine organisms. These unusual marginellids and *Cerithium* species, certainly warrants more research in the future.

Cerithium eburneum aliceae differs from the nominate subspecies in being a much thinner, more inflated shell, in having a longer and better developed siphonal canal, in lacking a nodulose sculpture pattern and being, instead, covered with fine, smooth spiral threads, and in lacking varices or axial ribs. The new subspecies is also much darker-colored than *C. eburneum eburneum,* with most specimens being almost solid dark brown.

Cerithium muscarum pacei new species (Plate 3, Figures 19, 20)

Material Examined: HOLOTYPE - Length 16 mm, width 7 mm, on algae in 1 ft. depth in enclosed salt water lake (the Fishing Pond) near Palmetto, Eleuthera Is., Bahamas, 1980 USNM 859822; PARATYPES - length 14 mm, same locality as holotype, USNM 859823; 15 specimens, same locality as holotype, Kevan Sunderland collection; 15 specimens, same locality as holotype, Robert Pace collection.

Description: Shell thin and fragile, very elongated; siphonal canal long, well developed, curved sharply to left; suture tightly impressed, producing scalariform appearance; whorls rounded but sharply-angled at mid-body; aperture oval; flaring; body whorl ornamented with 4 rows of large, oblong beads; beads connected by strong cord; fine spiral threads between 4 beaded cords; spire whorls with numerous fine spiral threads and 2 beaded cords; shell color dark chocolate brown with regularly-spaced axial flammules of darker blackish-brown; beaded cords of body whorl and spire whorls black; columella and siphonal canal white; protoconch and early whorls tan.

Etymology: Named for Mr. Robert Pace of Miami, Florida, who collected the holotype and type lot on Eleuthera.

Discussion: Like *Cerithium eburneum aliceae, C. muscarum pacei* is restricted to the salt water lake on Eleuthera Is., and represents an isolated population of the widespread Caribbean *C. muscarum* Say. *Cerithium muscarum pacei* differs from the nominate subspecies in being a much thinner, more elongated shell with a higher spire and more impressed suture, in having fewer rows of beads on the body whorl, and in having fewer rows of beads on the body whorl, and in being much darker colored. The solid dark-brown shell with the black beaded cords of *C. muscarum pacei* differs greatly from the gray shell with rows of brown specks of *C. muscarum muscarum.*

Family **Triviidae**
Genus *Trivia* Broderip, 1837
Subgenus *Niveria* Jousseaume, 1884

Trivia (Niveria) pacei new species (Plate 9, Figures 1, 2)

Material Examined: HOLOTYPE - Length 12 mm, width 9 mm, under coral rubble, 10 ft. depth, off southern end of South Bimini Is. (near Sunshine Inn Hotel), Bimini Chain, Bahamas, 1979, USNM 859830; PARATYPES - 4 specimens, same locality and depth as holotype, Robert Pace collection.

Description: Shell globose, terminals rounded, barely produced; dorsum with 12-16 ribs that radiate from deep, central, dorsal sulcus; smaller secondary ribs between primary ribs; base rounded, with 10-12 primary ribs on columella side and 12-14 primary ribs on labial side; aperture narrow at posterior end, widening at anterior end; fossula poorly-developed; shell color deep purple-black with paler purple terminals; interior of dorsal sulcus paler purple; ribs paler purple with black spots; base of shell often dark red.

Etymology: Named for Mr. Robert Pace of Miami, Florida, who collected the holotype on Bimini.

Discussion: *Trivia pacei* is closest to the widespread Caribbean *T. pediculus* (Linnaeus), but differs in being a deep purple-black shell instead of pink with black patches, by having the aperture widened at the anterior end instead of being narrow throughout, and by having fewer ribs on both the columellar and labial sides of the base. The spaces between the basal ribs are far wider than those of *T. pediculus,* giving *T. pacei* the appearance of a *Pusula* species when viewed from the base. *Trivia pacei* is also a consistently smaller species than *T. pediculus,* and in size, approaches the sympatric *T. suffusa* (Gray). *Trivia pacei* differs from *T. suffusa* by being a much more rounded, globose shell, by having a dark purple-black shell color instead of pink, in lacking the distinct fossula seen in the *T. suffusa,* and in having far fewer ribs on the base. To date, *T. pacei* is only known from the Bimini Chain of islands, and is probably endemic to the Bahamas.

Family **Muricidae**
Subfamily **Muricinae**
Genus *Chicoreus* Montfort, 1810

Chicoreus dunni new species (Plate 10, Figure 5)

Material Examined: HOLOTYPE - Length 36 mm, under dead coral slab, 1 m depth, southeastern end, Eleuthera Island, Bahamas, 1985, USNM 859835; PARATYPES - 15 specimens, same locality and depth as holotype, Kevan Sunderland collection.

Description: Shell small for genus, slender, elongated, with elevated spire and long siphonal canal; body whorl with 3 varices; intervarical regions with 1 large, rounded knob; varices with 6 highly fimbriated, long, slender, recurved spines; siphonal varices with 3 very elongated, slender, recurved, needle-like spines; body whorl with 6 highly fimbriated spiral cords; aperture large, oval; shell color uniformly very dark blackish-brown; edge of parietal shield orange; interior of aperture brown; spines of siphonal canal always longer than spines on varices.

Etymology: Named for the late Mr. Roger Dunn of Chicago, Illinois.

Discussion: *Chicoreus dunni* is closest to the Floridian and Gulf of Mexico *C. florifer* (Reeve) but differs in being a consistently smaller, more elongated shell with a higher spire and narrower body whorl, in having more delicate, needle-like spines, in having only three spines on the siphonal canal, and in being consistently solid blackish-brown in color. In general size, shape, and color, *C. dunni* looks very much like *C. trivialis* (Adams) from the southwest Pacific. *Chicoreus dunni,* in being slender and high-spired, also bears some resemblance to *C. cossmani* Abbott and Finlay from Jamaica. The Jamaican endemic, however, is much more heavily-built and higher-spired than *C. dunni* and has larger,more frondose spines. *Chicoreus cossmani* is also a more colorful shell, being pink, red, orange, yellow, and typically cream-tan colored. *Chicoreus dunni* is also similar to *C. pudicolor* (Reeve) from the Netherlands Antilles, but differs in having better-developed and longer spines, a longer siphonal canal, and by having three siphonal spines instead of two.

Chicoreus dunni occurs in the same habitat as does another Bahamian endemic, *Conus jucundus* Sowerby (= *C. abbotti* Clench). Both species live under large slabs of dead coral on top of a hard limestone substrate in shallow water (approx. 1 m depth). Both *Chicoreus dunni* and *Conus jucundus* are relatively plentiful off Eleuthera Island. Specimens of a black *"Chicoreus florifer"* reported from Bimini may prove to be *C. dunni.* To date, *C. dunni* has only been collected off Eleuthera and is probably endemic to the Bahamas Platform.

Subfamily **Muricopsinae**
Genus *Muricopsis* Bucquoy, Dautzenberg, and Dollfuss, 1882

Muricopsis sunderlandi new species (Plate 9, Figures 3, 4, and 5)

Material Examined: HOLOTYPE - Length 13 mm, width 8 mm, under coral rubble, 10 m depth off Cay Sal, Cay Sal Bank, Bahamas, 1982, USNM 859848; PARATYPES - length 14 mm, same locality and depth as holotype, USNM 859849; 15 specimens, same locality as holotype, Kevan Sunderland collection.

Description: Shell stocky, inflated, with spire lower than average for the genus; shell with 6 varices per whorl; varices extremely ornate, with body whorl varices having 4 long, tubular spines per varix; 1 or 2 small scale-like projections between 4 large spines; varices on siphonal canal with 2 large spines and several small spines in between; spire whorl varices with 1 large spine and 3 small spines per varix; shell color bright red-orange with single thin, white band around mid-body; siphonal canal white; intervarical areas smooth, with 4 large cords; mid-body cord corresponds to white band; columella white; interior of aperture red-orange.

Etymology: Named for Mr. Kevan Sunderland, who collected the type lot on Cay Sal.

Discussion: *Muricopsis sunderlandi* is a spectacular addition to the western Atlantic muricid fauna. The consistent bright orange-red color and long needle-like spines make *M. sunderlandi* a candidate for the most beautiful Caribbean murex and set it aside from any other known Caribbean muricopsines. The closest Atlantic relative of *M. sunderlandi* is *M. oxytatus* (M. Smith), but the new species differs from that common, widespread Caribbean species in being a smaller, more delicate shell with a much stockier, rounder body form, in having a lower spire, in having longer, more delicate spines, and in being consistently bright orange-red with a white band and white siphonal canal. *Muricopsis oxytatus* can come in a variety of colors, from white, pink, lilac, to orange or brown-mottled, but it never exhibits the intensity of color seen in *M. sunderlandi.*

Muricopsis sunderlandi appears to be endemic to the Bahamas, and is most common on Cay Sal Bank. Mr. William Lyons (Florida State Dept. of Natural Resources) has a similar large, extra-spiny specimen of *Muricopsis* taken by SCUBA divers from a sealab that was submerged off Grand Bahama Island. This large, pink-spined specimen may be referable to *M. sunderlandi.*

Family **Conidae**

Genus *Conus* Linnaeus, 1758

Conus leekremeri new species (Plate 9, Figures 7, 8)

Material Examined: HOLOTYPE - Length 30 mm, width 9 mm, brought up in deep water lobster trap, from 240 m depth off southern coast of Grand Bahama Island, Bahamas, 1979, USNM 859885; PARATYPE - 1 specimen, same locality and depth as holotype, Robert Pace collection.

Description: Shell thin, elongated, with high, widely scalariform spire; body whorl shiny, highly polished, ornamented with numerous low, rounded, spiral cords; cords become large and more numerous around the anterior end; spire whorls ornamented with 4 large, spiral threads; spire whorls concave, producing canaliculated whorls; canaliculate spire edged with low, rounded carina along edge of shoulder; carina follows edge of suture on canaliculate spire whorls; sides of body whorl distinctly concave and indented, producing an emaciated, waisted appearance; aperture long and very narrow; body whorl, spire, and interior of aperture pure white; periostracum thin, transparent yellow.

Etymology: Named for Lee Kremer of Crystal Lake, Illinois, a dedicated student of the *Conidae.*

Discussion: *Conus leekremeri* is closest to the widespread western Atlantic *C. villepini* Fischer and Bernardi, but differs in being a smaller, more elongated shell with much more concave sides, and by having a lower,and distinctly canaliculate, spire. *Conus villepini* is also a more colorful shell, with many large brown flammules, while *C. leekremeri* is pure white. The common, shallow water *C. emaciatus* Reeve from the southwestern Pacific is the only other known cone shell that has sides as indented as those of *C. leekremeri.* This unusual new deep water species lives with the other Bahamian endemics, *C. lindae* Petuch and *C. pacei* Petuch, and such other rare, deep water species as *Volutifusus piraticus* (Clench and Aguayo) and *Oliva bahamasensis* Petuch and Sargent.

Conus lindae new species (Plate 9, Figures 9, 10)

Material Examined: HOLOTYPE - Length 31 mm, width 17 mm, brought up in deep water lobster trap from 240 m depth off southern coast of Grand Bahama Is., Bahamas, 1979, USNM 859886; PARATYPES - 1 specimen same locality and depth as holotype, Charles Key collection; 1 specimen, trawled from 250 m depth in Straits of Florida, west of Bimini, Bahamas, 1963, by *R/V Gerda,* invertebrate museum collection of RSMAS, University of Miami.

Description: Shell solid, stocky, broad across shoulder, with relatively low spire; shoulder distinctly rounded; spire whorls slightly canaliculate; body whorl shiny, highly polished, with waxy feel; aperture narrow; protoconch very large, rounded, mamillate; shell color varying from pure white (paratype, Key collection) to pale pink, overlaid with 2 bands of salmon-pink dots (holotype); wide bands of salmon-pink blotches arranged with one above mid-body and one below mid-body near anterior end; rounded shoulder and canaliculate spire whorls marked with evenly-spaced, thin, radiating, pale salmon-orange flammules; interior of aperture white; periostracum thin, smooth, yellow, translucent.

Etymology: Named for my wife, Linda J. Petuch.

Discussion: *Conus lindae* does not closely resemble any known cone species from the western Atlantic. It does, however, closely resemble some pale varieties of *C. angasi* Tryon, from deep water off eastern Australia. This similar appearance in shell form and color indicates that, along with *C. lightbourni* Petuch (Plate 6, Figures 7, 8, 9) from deep water off Bermuda, *C. lindae* represents the second-known western Atlantic species of the *Floraconus* Iredale species complex. *Conus lindae* might, at first glance, be confused with the unrelated *C. villepini* Fischer and Bernardi, but the new species differs in being a smaller, stockier shell with a lower, canaliculate spire, and by having distinctly rounded shoulders. The color pattern of salmon-pink blotches and dots also easily separates *C. lindae* from any known variety of *C. villepini.*

Conus pacei new species (Plate 9, Figures 11, 12; Plate 7, Figure 6)

Material Examined: HOLOTYPE - Length 19 mm, width 8 mm, taken in baited lobster trap, from 250 m depth, off southern coast of Grand Bahama Island, Bahamas, 1977, USNM 859888; PARATYPES - length 20 mm, same depth, locality, and date as holotype, Robert Pace collection; length 18 mm, same depth and locality as holotype, Charles Key collection; 1 specimen, Gene Everson collection.

Description: Shell thin, fragile, delicate, elongated in shape; spire high, protracted; shoulder sharp-angled, smooth; protoconch large, 2 whorls; aperture narrow; body whorl ornamented with numerous incised spiral sulci, giving shell silky appearance; spire whorls ornamented with 3-4 thin, spiral threads; shell color pure white; interior of aperture white; periostracum thin, yellow, translucent.

Etymology: Named for Mr. Robert Pace of Miami, Florida, in recognition of his many discoveries of new Caribbean mollusks.

Discussion: *Conus pacei* is the newest member of a close-knit species complex of deep water cones comprising *Conus mazei* Deshayes, *C. rainseae* McGinty, and *C. macgintyi* Pilsbry. In having a rough-textured shell, *C. pacei* is closest to *C. macgintyi* Pilsbry, but differs in being a much smaller, stockier shell, by lacking the beaded shoulder and spire whorls, and by being a pure white shell instead of having bands of large brown spots and brown spire flammules. In shape and size, *C. pacei* comes closest to *C. rainseae* McGinty from off Yucatan, Mexico, but differs in having a strongly sculptured body whorl and by lacking a color pattern. From the southern Caribbean *C. mazei* Deshayes, *C. pacei* differs in being a smaller, stockier shell, by having a much higher spire, by having a sculptured body whorl, and by lacking a spotted color pattern.

In many ways, particularly in size, proportions, and sculpturing, *C. pacei* resembles a pale *C. insculptus* Kiener from deep water in the Melanesian-Philippines region. This new deep water Bahamian species also resembles the Caribbean Pliocene fossil species *C. gracilissimus* Guppy, but differs in lacking the row of beads along the shoulder. *Conus gracilissimus* may be the direct ancestor of *C. pacei. Conus rainseae, C. pacei, C. mazei,* and *C. macgintyi* are shown together on Plate 7 for easy comparison.

Family **Terebridae**
Genus *Terebra* Bruguière, 1789
Subgenus *Strioterebrum* Sacco, 1891

Terebra (Strioterebrum) biminiensis new species (Plate 9, Figures 15, 16)

Material Examined: HOLOTYPE - Length 33 mm, width 7 mm, on sand bottom 1 m depth, Nixon's Harbour, South Bimini Island, Bimini Chain, Bahamas, 1975, USNM 859893; PARATYPE - length 22 mm, on beach, Nixon's Harbour, South Bimini Island, Robert Pace collection.

Description: Shell slender, elongated, with straight-sided whorls; whorls with 21-25 narrow, rounded axial ribs; ribs slightly curved, being bent toward aperture near suture, and straight on main body of whorl; deeply impressed sulcus below suture, producing wide subsutural band; subsutural band intersected by axial ribs, producing beaded effect; body whorl, subsutural band, and spire whorls covered by numerous, fine, raised spiral threads; aperture long and narrow, produced anteriorward; shell color uniform pale pinkish tan; interior of aperture rose-pink with tan overtones.

Etymology: Named for the Bimini Chain of islands, the Bahamas, the type locality.

Discussion: *Terebra biminiensis* is the newest member of the *T. protexta* Conrad species complex, which includes several similar-appearing, small species that are distributed throughout the western Atlantic. *Terebra biminiensis* is closest to *T. protexta,* but differs in being a much larger, more elongated shell, in having a better-developed and more pronounced subsutural band, in having more axial ribs, in having a more sloping last whorl and more elongated aperture, and in being a pale pinkish-tan color instead of dark tan or brownish-purple. *Terebra biminiensis* appears to be the largest species in the *T. protexta* complex, and with its prominent subsutural band, bears resemblance to a slender *T. dislocata* Say. To date, *T. biminiensis* has only been found in the Bimini Chain of islands, but may possibly be found at other localities in the Bahamas.

Terebra (Strioterebrum) pacei new species (Plate 9, Figures 17, 18)

Material Examined: HOLOTYPE - Length 24 mm, width 5 mm, in sand, 1 m depth, Nixon's Harbour, South Bimini Island, Bimini Chain, Bahamas, 1975, USNM 859894; PARATYPE - Length 22 mm, on beach, Nixon's Harbour, South Bimini Island, Bahamas, Robert Pace collection.

Description: Shell elongated, with distinctly concave whorls; last whorl elongated, with rounded and tapered base; aperture elongated; whorls ornamented with 21-24 large, evenly-spaced axial ribs; subsutural band well-developed, projecting beyond outline of shell, producing concave appearance to whorls; axial ribs extend onto subsutural band, producing evenly-spaced, oval-shaped beads; axial ribs overlaid with numerous large, raised spiral threads; subsutural beads smooth, without threads; shell color pale lilac purple with scattered, large reddish-brown axial flames; base of shell darker purple; columella and siphonal canal brown; protoconch and early whorls dark purple-brown; interior of aperture white.

Etymology: Named for Mr. Robert Pace of Miami, Florida, in recognition of his extensive exploratory collecting in the Bimini area.

Discussion: *Terebra pacei* is a distinctive little species, combining the features of several small *Terebra* species from the Caribbean region. *Terebra pacei* is closest to *T. concava* Say from the North American mainland, but differs in being a larger, more slender shell with a much more elongated aperture, in having prominent axial ribs, in having a much more elongated last whorl with a rounded base instead of an angled one, and in having prominent spiral threads superimposed on top of the axial ribs. The color patterns also differ greatly; *T. pacei* is a lilac-colored shell with scattered orange-tan axial flames and dark purple early whorls and base, while *T. concava* is uniformly pale yellow or gray.

Terebra pacei may be endemic to the Bahamas, although similar-appearing, large specimens have recently been dredged from 200 ft. depths off Palm Beach. Further study may show them to be conspecific.

PLATE 9. NEW GASTROPODS FROM THE BAHAMAS AND COZUMEL ISLAND

Figures 1, 2: *Trivia (Niveria) pacei* Petuch, n.sp., dorsal and ventral views of holotype, 12 mm.

Figures 3, 4: *Muricopsis sunderlandi* Petuch, n.sp., dorsal and ventral views of holotype, 13 mm.

Figure 5: *Muricopsis sunderlandi* Petuch, n.sp., dorsal view of paratype.

Figure 6: *Conus sphacelatus* Sowerby, 1833, ventral view of 17 mm specimen, Grand Cayman Island, for comparison with *Conus kirkandersi* Petuch.

Figures 7, 8: *Conus leekremeri* Petuch, n.sp., dorsal and ventral views of holotype, 30 mm.

Figures 9, 10: *Conus lindae* Petuch, n.sp., dorsal and ventral views of holotype, 31 mm.

Figures 11, 12: *Conus pacei* Petuch, n.sp., dorsal and ventral views of holotype, 19 mm.

Figures 13, 14: *Conus kirkandersi* Petuch, n.sp., dorsal and ventral views of holotype, 15 mm.

Figures 15, 16: *Terebra (Strioterebrum) biminiensis* Petuch, n.sp., dorsal and ventral views of holotype, 33 mm.

Figures 17, 18: *Terebra (Strioterebrum) pacei* Petuch, n.sp., dorsal and ventral views of holotype, 24 mm.

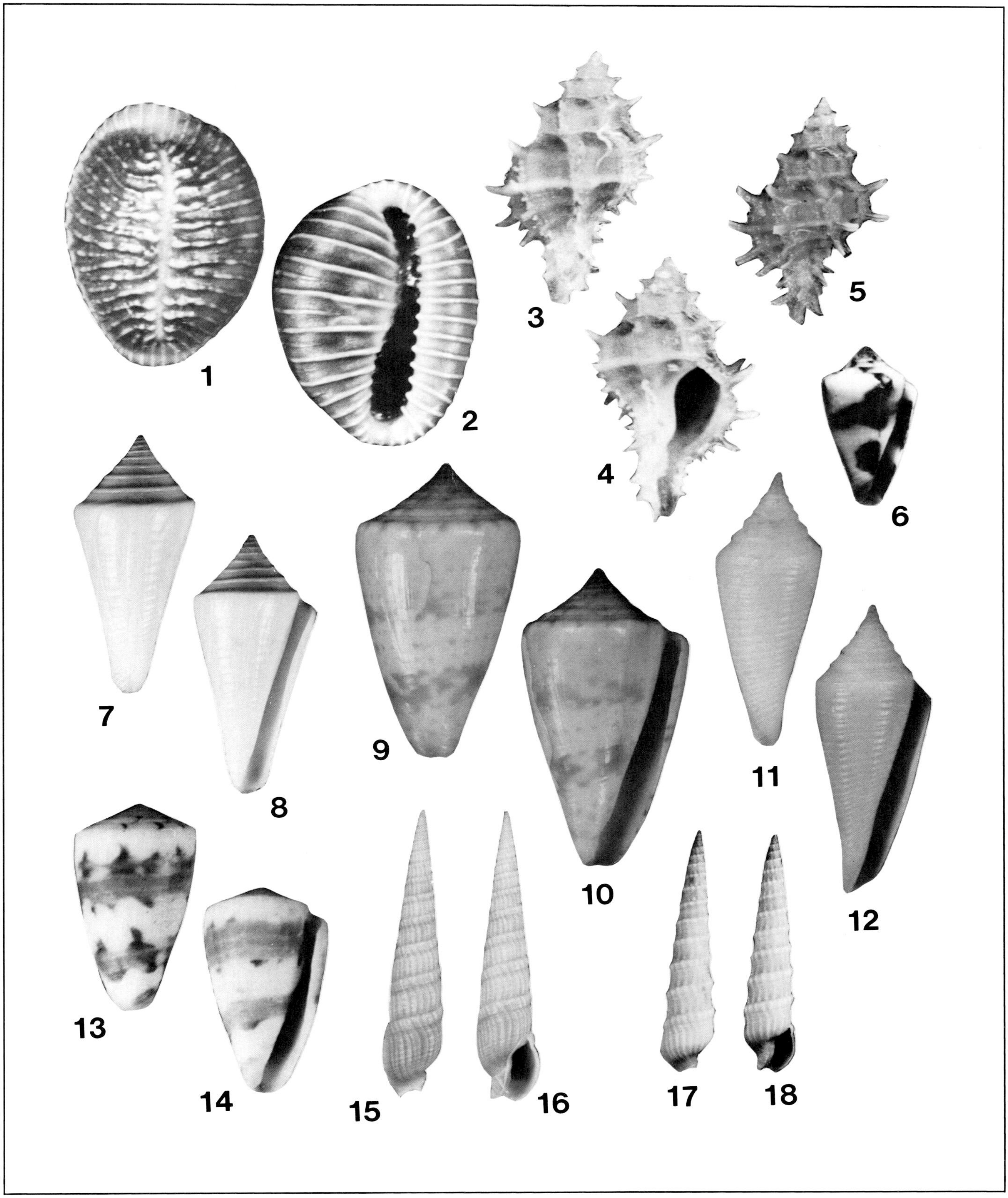
1
2
3
4
5
6
7
8
9
10
11
12
13
14
15
16
17
18

PLATE 10. NEW AND SELDOM-SEEN GASTROPODS FROM FLORIDA, THE BAHAMAS, AND HONDURAS.

Figures 1, 2, 3: *Architectonica sunderlandi* Petuch, n.sp., dorsal, ventral, and lateral views of holotype, 20 mm diameter.

Figure 4: *Architectonica sunderlandi* Petuch, n.sp., lateral view of 31 mm diameter paratype.

Figure 5: *Chicoreus dunni* Petuch, n.sp., ventral view of holotype, 36 mm.

Figures 6, 7: *Zafrona taylorae* Petuch, n.sp., dorsal and ventral views of holotype, 11 mm.

Figures 8, 9: *Murexiella taylorae* Petuch, n.sp., dorsal and ventral views of holotype, 16 mm.

Figures 10, 11: *Pterynotus radwini* Harasewych and Jensen, 1979, holotype, 31 mm, 300 m depth off Belize.

Figures 12, 13: *Conus eversoni* Petuch, n.sp., dorsal and ventral views of holotype, 18 mm.

Figure 14: *Conus eversoni* Petuch, n.sp., dorsal view of paratype (Sunderland collection), Utila Island, Honduras.

Figure 15: *Fulgurofusus xenismatis* Harasewych, 1983, 33 mm paratype, 200 m depth, off NE Nicaragua.

Figures 16, 17: *Cerithioclava garciai* Houbrick, 1985, Utila Island, Honduras, 65 mm.

Figures 18, 19: *Conus commodus* A. Adams, 1854, dorsal and ventral views of 18 mm specimen, off Roatan Island, Honduras.

Figures 20, 21: *Zafrona sunderlandi* Petuch, n.sp., dorsal and ventral views of holotype, 10 mm.

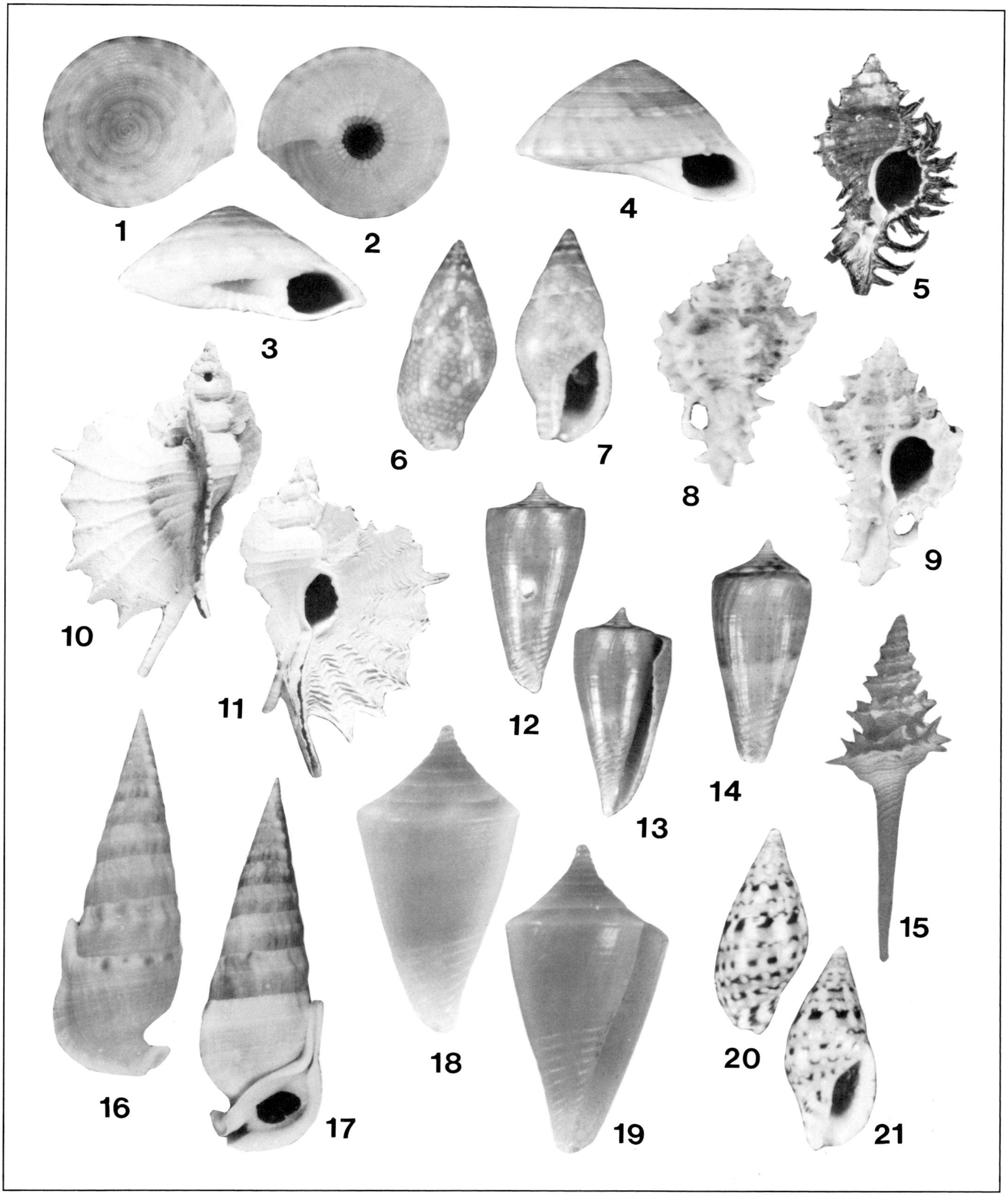
1
2
3
4
5
6
7
8
9
10
11
12
13
14
15
16
17
18
19
20
21

CHAPTER 5

New Discoveries from Honduras and Eastern Central America

The Gulf of Honduras area, particularly around the Bay Islands of Roatan, Utila, and Guanaja, contains one of the most unexplored regions of the Caribbean. Since the late 1960's, collections taken from shrimpers working the upwelling systems of this area have shown that the Gulf region contains a relict pocket similar to the one off western Florida and northern South America (PETUCH 1981A; 1982). Such supposedly-extinct genera as *Pleioptygma (Pleioptygma helenae* (Radwin and Bibby)), *Cerithioclava (Cerithioclava garciai* Houbrick), and *Falsilyria (Falsilyria morrisoni* Petuch), have been taken rather frequently off the coast of Roatan Island. Along with the first-known Caribbean species of *Bolma* and *Agaronia* (described here), these last-mentioned species give the Honduran fauna an archaic look.

Of special interest from this area is a species radiation of the volutid genus *Falsilyria* that is unique in the Caribbean. This ancient group, which first appears in the early Miocene fossil beds of Florida, has survived along the eastern coasts of Honduras and Nicaragua and has speciated into a number of interesting forms such as *Falsilyria morrisoni* Petuch (Plate 15, Figures 3, 4), *F. kotorai* Petuch (Plate 14, Figures 3, 4), *F. garciai* Petuch (Plate 14, Figures 5, 6), *F. demarcoi* (Olsson) (Plate 13, Figures 5, 6), *F. retemirabilis* Petuch (Plate 14, Figures 7, 9), and three new species which are described here. The rare *Lyria (Enaeta) reevei* (Plate 17, Figures 18, 19) has also recently been collected.

Also trawled along this coast are the endemic muricid *Pterynotus radwini* Harasewych and Jensen (Plate 10, Figures 10, 11), the spiny turbinellid *Histricosceptrum xenismatis* Harasewych, and the cones *Conus lorenzianus* Dillwyn, *Conus garciai* daMotta, and the lost species *Conus commodus* Adams (Plate 10, Figures 18, 19). SCUBA divers working the reefs around the Bay Islands have uncovered a new cone shell fauna that includes *Conus kulkulcan* Petuch (Plate 15, Figures 12, 13, 14) and four new species that are described here. Farther north along the Yucatan coast and Cozumel Island, the large, yellow olive *Oliva maya* Petuch and Sargent (Plate 16, Figures 7, 8), a new reef cone *(Conus kirkandersi* Petuch, described here), and the large lost cowrie, *Cypraea aubryana* Jousseaume (Plate 6, Figures 1, 2, 3, 4; Plate 22, Figures 1, 2), have recently been collected. This latter species is discussed in detail later in this chapter.

New Species From Honduras And Eastern Central America

Bolma (Senobolma) sunderlandi new species
Distorsio robinsoni new species
Chicoreus emilyae new species
Murex garciai new species
Murex samui new species
Zafrona sunderlandi new species
Ancilla (Amalda) pacei new species
Agaronia hilli new species
Agaronia leonardhilli new species

Dibaphimitra janetae new species
Falsilyria harasewychi new species
Falsilyria hilli new species
Falsilyria sunderlandi new species
Voluta lindae new species
Conus kirkandersi new species
Conus eversoni new species
Conus harlandi new species
Conus magnottei new species
Conus sunderlandi new species
Hindsiclava tippetti new species

Family **Turbinidae**
Subfamily **Astraeinae**
Genus *Bolma* Risso, 1826
Subgenus *Senobolma* Okutani, 1964

Bolma (Senobolma) sunderlandi new species (Plate 11, Figures 1, 2)

Material Examined: HOLOTYPE - Height 16 mm, width 15 mm, trawled by commercial shrimpers from 35 m depth off Roatan Is., Honduras, 1977, USNM 859814; PARATYPES - height 16 mm, same locality and depth as holotype, Kevan Sunderland collection; 1 specimen, Roatan Is., Robert Pace collection.

Description: Shell thick, solid, turbinate in shape; spire elevated with stepped appearance; shoulder very sharply angled, carinated; area between shoulder carina and suture flattened, distinctly tabulate; body whorls and spire whorls sculptured with numerous fine, smooth, spiral threads; outer edge of lip thickened, flaring; columella and base well-developed, white in color; interior and periphery of aperture nacreous; shell color pale straw-tan; operculum oval, smooth, shelly, white in color.

Etymology: Named for Kevan Sunderland, Fort Lauderdale, Florida.

Discussion: The discovery of an American *Bolma* is of great importance since, according to BEU and PONDER (1979, p. 6), there have been no records of the genus from the Americas, either living or as a fossil. Even more interesting is the fact that *B. sunderlandi* belongs to the subgenus *Senobolma,* which was previously thought to be restricted to Japan. The new Caribbean species is closest to *B. (Senobolma) venusta* (Okutani), but differs in lacking short peripheral spines on the early whorls, by having a stronger shoulder angle and tabulate spire, and by having a pale tan color instead of reddish-brown. The spiral sculpture on *B. sunderlandi* is also finer than that of the Japanese *B. venusta.*

Superfamily **Cypraecea**
Family **Cypraeidae**
Genus *Cypraea* Linnaeus, 1758

Cypraea aubryana Jousseaume, 1869 (Plate 6, Figures 1, 2, 3, 4; Plate 22. Figures 1, 2)

Material Examined: Length 45 mm, trawled by *R/V John Elliot Pillsbury,* University of Miami, from 120 m depth in channel between Cozumel Island and main land of Yucatan, Mexico (station P-

598), 1966, DMNH 164799.

Discussion: Previously known from only two specimens, the 48 mm holotype from Guadeloupe and a 49 mm specimen from St. Vincent, *Cypraea aubryana* is amongst the rarest of the cowrie shells. The third known specimen, taken off Cozumel Is., Mexico and illustrated here, demonstrates the fact that the species is widespread throughout the Caribbean Sea. The shell is so distinctive that it is amazing that only three specimens have been recognized.

Most recent workers, including TAYLOR and WALLS (1975) and WAGNER and ABBOTT (1978), have regarded *C. aubryana* as a synonym of *C. surinamensis* Perry, 1811. EMERSON and OLD (1965) correctly stated that *C. aubryana* was distinct from *C. surinamensis,* but presumed that it was not a Caribbean species and that Jousseaume's type locality of Guadeloupe was erroneous. Interestingly enough, Mr. BURGESS, in both of his definitive volumes (1970), (1985), fails to mention the taxon *aubryana* as either a synonym or otherwise. This is particularly surprising since Jousseaume described the new taxon in a well-known publication series with the word *Cypraea* in the title, and that the work was subsequently discussed in detail by EMERSON and OLD in *The Nautilus* (1965).

Morphologically, *C. aubryana* is very different from *C. surinamensis.* Besides being a larger shell (av. 47 mm as opposed to 35 mm in *C. surinamensis),* it has a much more inflated outline and lacks the well-developed, projecting extremities that are characteristic of *C. surinamensis.* In *C. surinamensis,* the labial and columellar dentition extend onto the base of the shell. The dentition is especially well-developed on the labial region, where the teeth often extend across the entire base, especially at the anterior and posterior ends. In *C. aubryana,* the teeth are much smaller and finer, and do not extend onto the base. *Cypraea aubryana* also has fewer teeth than a comparable-sized *C. surinamensis.* The specimen of *C. aubryana* illustrated here (45 mm in length) has 22 columellar teeth and 27 labial teeth, while the *C. surinamensis* illustrated by BAYER (1971) (37 mm in length and large for the species) has 24 columellar and 28 labial teeth.

The *C. aubryana* illustrated here was dead - taken, drilled by *Cliona* sponges, and had lost most of its color. Still the dorsum is a pale yellow-orange with a darker central blotch, and the base and margins are pale orange. The 49 mm specimen of *C. aubryana* from St. Vincent Is., Lesser Antilles that was illustrated in color by SOWERBY (1880, plate 37, figures 529-530), is the best representation of the species to date. He shows a brilliant orange shell covered with large blotches and speckles of dark brown, similar to *C. surinamensis.* In many ways, the color pattern is reminiscent of *C. coronata* Schilder, 1930 from South Africa. The inflated body form, fine teeth, and lack of extremities also point to a possible relationship to the subgenus *Schilderia* (as exemplified by *C. achatidea).* Photographs of the *C. aubryana* illustrated by Sowerby are shown here on Plate 6, Figure 1, 2 along with the Cozumel specimen.

Superfamily **Tonnacea**
Family **Cymatiidae**
Genus *Distorsio* Röding, 1798
Subgenus *Rhysema* Clench and Turner, 1957

Distorsio (Rhysema) robinsoni new species (Plate 11, Figures 3, 4)

Material Examined: HOLOTYPE - Length 53 mm, width 33 mm, on sand bar,low tide, Oak Ridge, Roatan Is., Honduras, 1982, USNM 859832; PARATYPES - length 48 mm, (with periostracum), trawled by fishermen from 50 m depth off Salvador, Bahia State, Brazil, 1985, USNM 859833; length 50 mm, off Roatan Is., Honduras, collection of Kevan Sunderland; 3 specimens from same locality and date as holotype, collection of David Robinson.

Description: Shell thin but strong, lightweight; body whorl inflated, with large, distorted swelling after each varix; spire low for subgenus; body whorl and spire whorls ornamented with coarse reticulated pattern produced by intersection of thick axial and spiral cords; junction of cords marked by rounded bead; parietal shield expansive, with reticulated pattern under glossy layer of enamel; outer lip flaring, flattened, with 8 large, pointed serrations along edge; anterior most serrations largest, bladelike; outer edge of lip straight; posterior most edge of lip squared-off in sharp-edged right angle; columella greatly twisted, producing indentation; anterior area of columella below indentation straight, with 12-14 small teeth; posterior area of columella above indentation with 3 large, elongated, sloping teeth and one large central tooth; labial teeth extend across aperture; on body whorl, beaded cord below suture thicker than other cords; subsutural cord separated from next cord by wide space; space widest on dorsal hump; other cords on body whorl equally-spaced; body whorl color white with pale orange, wide, mid-body band; early whorls orange; parietal shield and columella orange to tannish-orange with white reticulated pattern; outer lip white with orange-tan radiating bands between serrations; periostracum yellow tan in color, adherent, fuzzy and velvety, with only few scattered long hairs on edges of varices; siphonal canal wide, straight, tipped with dark orange-brown color.

Etymology: Named for David Robinson, Tulane University, New Orleans, who collected the holotype and kindly donated material for study.

Discussion: Finding a new *Distorsio* species in the Caribbean is, indeed, an exiting discovery. The fact that the species is found both in Honduras and Brazil shows that it is widespread throughout the western Atlantic. *Distorsio robinsoni* may possibly have been collected before, but was probably misidentified as the closely-related species *D. clathrata* (Lamarck). The two species probably coexist throughout the Caribbean region. *Distorsio robinsoni* differs from *D. clathrata* in a number of consistent ways; by having a lower spire, wider shoulder, and more inflated body whorl; by having a straight siphonal canal and anterior columella instead of a twisted one; by having a sharp, right-angled posterior edge to the lip and varices instead of a rounded one; by having larger, projecting serrations on the anterior of the lip edge; by having a larger medial tooth on the inner edge of the lip; by having a larger indentation into the columella; by having,on the dorsal hump, a wide gap between the subsutural cord and the next cord; by having a wide, pale orange band around the mid-body; and by having a straight edge to the outer lip instead of a rounded one. The periostraca of the two species also differ greatly. The periostracum of *D. robinsoni* is yellow-tan, velvety, and virtually devoid of hairs except for a few on the edge of the varix. The periostracum of *D. clathrata,* on the other hand, is dark brown or gray and profusely covered with long, hair-like bristles.

In many ways, *D. robinsoni* resembles a short-spired version of the Panamic-Eastern Pacific *D. decussata* (Valenciennes). The wide gap between the subsutural and shoulder cords is seen in both species. Both *D. decussata* and *D. robinsoni* appear to have evolved from a common, widespread, Pliocene fossil species, *D. gatunensis* Toula.

Superfamily **Muricacea**
Family **Muricidae**
Genus *Chicoreus* Montfort, 1810

Chicoreus emilyae new species (Plate 13, Figures 1, 2)

Material Examined: HOLOTYPE - Length 55 mm, width 45 mm, trawled by commercial shrimpers from 20 m depth off Punta Patuca, Honduras, 1977, USNM 859834; PARATYPES - 3 specimens, Axel A. Olsson collection, RSMAS, University of Miami, off Bocas de Toro, Panama; 3 specimens, on coral rubble bottom, 60 ft. depth by SCUBA off Roatan Is., Honduras, Robert Pace collection.

Description: Shell rotund, inflated, wide across shoulder; 3 varices per whorl; 1 heavily-knobbed axial rib between varices; 2 knobs along shoulder largest, sharply-angled; varix with 6 side, flattened, highly frondose and branching spines; varices of siphonal canal with 5 long, flattened, and frondose spines; spines of siphonal canal and anteriormost varical spines often fusing into flattened mass of interconnected fronds; body whorl and spire whorls sculptured with numerous fine, raised spiral threads; frondose spines of varices secondarily sculptured with numerous fine, raised spiral threads; frondose spines of varices secondarily sculptured with numerous large scales; aperture round, very large in proportion to shell size; posterior anal canal large, open; siphonal canal short in proportion to shell size; parietal shield narrow, with edge unattached to shell; color varying from orangish-tan to dark brown, overlaid with fine, dark brown lines.

Etymology: Named for Dr. Emily Vokes of Tulane University, New Orleans, foremost authority on the *Muricidae*.

Discussion: *Chicoreus emilyae* is closest to *C. florifer* (Reeve) from Florida, the Bahamas, and Cuba, but differs in being a much rounder, more globose shell, in having a wider aperture, in having a lower spire and shorter siphonal canal, and in having five large frondose spines on the siphonal canal instead of four. The varical spines of *C. emilyae* are also larger, broader, more flaring, and have more well-developed side branchings than those of *C. florifer.* Particularly distinctive are the long, ramose siphonal canal spines of *C. emilyae,* which interconnect and overlap to produce a winglike varix. *Chicoreus emilyae* ranges from Roatan Is. south to Panama, and is the southern representative of the *C. florifer-dilectus-rachelcarsonae* species complex of the Florida-Bahamas region.

Genus *Murex* Linnaeus, 1758

Murex garciai new species (Plate 11, Figures 9, 10)

Material Examined: HOLOTYPE - Length 45 mm, width 24 mm, trawled by commercial shrimpers from 35 m depth off northern coast of Roatan Island, Bay Islands, Honduras, 1979, USNM 859839; PARATYPES - 1 specimen, trawled off Roatan Is., Honduras, Robert Pace collection; 2 specimens, trawled off Roatan Is., Honduras, Emilio Garcia collection; 4 specimens, no locality data, Kevan Sunderland collection.

Description: Shell broad, inflated, thick, with angled shoulder; whorls and varices slope gradually to siphonal canal, producing triangular-shaped varices; 3 varices per whorl; spire moderately elevated; intervarical areas with 3 bead-covered axial ribs; varices with numerous ridges but with only 1 large spine at shoulder; body whorl with 13-16 raised spiral cords; spiral cords correspond to ridges and shoulder spines on varices; siphonal canal with 1 short spine; shell color varying from pale gray to grayish-brown with 2 prominent bands of dark brown or black; bands darker on varices, producing characteristic dark, varical checkers; interior of aperture white, with 2 dark brown bands showing through; some spiral cords on body whorl marked with rows of tiny, elongated dark brown dots.

Etymology: Named for Dr. Emilio Garcia of Lafayette, Louisiana, who collected the holotype.

Discussion: *Murex garciai* is closest to the southern Caribbean *M. messorius* Sowerby, but differs in being less rounded and inflated, with angled, triangular varices, by lacking a distinct constriction at the junction of the body whorl and siphonal canal, in having a finer, more ornate ribbing pattern, and in having the two dark bands on the body and the two dark checker-like patches on each varix. *Murex messorius* often has two or three spines on the siphonal canal, while *M. garciai* has only one. At present, *M. garciai* has only been collected off Roatan Is., Honduras.

Murex samui new species (Plate 11, Figures 7, 8)

Material Examined: HOLOTYPE - Length 76 mm, width 35 mm, in 3 m depth in Turtle Grass *(Thalassia)* bed off Monkey River, Belize, 1979, USNM 859840.

Description: Shell large, inflated, with bulbous body whorl; 3 varices per whorl; body whorl and varices ornamented with 20 raised, spiral cords; cords on varices become rib-like, often ending in small, scale-like spine; varix with 3 large spines, 1 at shoulder, 1 at mid-body, and 1 at base; short, stubby, secondary spines between 3 large spines on some varices; shoulders of spire varices with 1 large projecting spine; siphonal canal with 1 large spine near top, several smaller scale-like spines below; end of siphonal canal smooth; columellar region with large, erect parietal shield that does not adhere to body of shell; aperture very large and flaring, round in shape; shell color uniformly pure white (holotype with some patches of rust stains); intervarical regions with 3 weak, knobbed, axial ribs.

Etymology: Named for the late Mr. Stephen Samu of Palm Beach, Florida, who collected the holotype. Before his untimely death, Mr. Samu showed me several more specimens of the new murex, all identical to the holotype, that were in his private collection. Unfortunately, the collection has since disappeared, along with the potential paratypes of this new species.

Discussion: *Murex samui* is one of the largest of the northern Caribbean *Murex* s.s. species. This new shallow water species is closest to *M. woodringi* Clench and Farfante from the West Indies and northern Colombia, but differs in being a larger, more inflated shell with a larger aperture, in having coarser sculpturing with only three intervarical axial ribs, by having larger beadlike knobs on the intervarical ribs, and in having larger and more numerous spines on the varices. *Murex woodringi* is a dark gray or brown shell, while *M. samui* differs greatly in being pure white. At present, *M. samui* is only known from the Turtle Grass beds along the southern coast of Belize, but it may extend farther south into the Gulf of Honduras.

Family **Columbellidae**
Genus *Zafrona* Iredale

Zafrona sunderlandi new species (Plate 10, Figures 20, 21)

Material Examined: HOLOTYPE Length 10 mm, under slab of dead coral, 3 m depth off southern coast of Utila Is., Bay Islands, Honduras, 1986, USNM 859905; PARATYPES - 5 specimens, same locality and depth as holotype, Sunderland collection.

Description: Shell fusiform, smooth, polished; aperture one-half length of shell; columella with raised, lipped callus; inside of lip with 8-10 denticles; columella with 1 large plication on inside of aperture; color bright orange-tan overlaid with 4 bands (on body whorl) of evenly-spaced, alternating, dark brown and white dots and dashes in a chainlike pattern; edge of suture bordered with large white and dark brown checkers; anterior tip of shell marked with alternating white and dark brown flammules; spire whorls with 1 chainlike band of of alternating dark brown and white dots and subsutural band of large white and dark brown checkers; interior of aperture and columella white; protoconch and postnuclear whorls tan; early whorls with very faint, weak axial riblets.

Etymology: Named for Mr. Kevan Sunderland of Ft. Lauderdale, Florida, who collected the type lot.

Discussion: *Zafrona sunderlandi* is closest to *Zafrona pulchella* (Blainville), both in size and shape, but differs in being a smoother shell, lacking axial riblets on the body whorl, and in having a color pattern made up of chainlike bands of brown and white dashes on a bright orange-tan background. The chainlike color pattern is reminiscent of *Mitrella* species, but the retention of the protoconch into adulthood, the single large columellar plication, and the raised columellar lip of *Z. sunderlandi* point to its inclusion in *Zafrona*.

The three species of northern Caribbean *Zafrona*, *Z. pulchella* (Blainville), *Z. taylorae* Petuch, and *Z. sunderlandi* Petuch, all have very different ecologies and habitat preferences. *Zafrona pulchella* lives under rocks and rubble in the intertidal zone; *Z. taylorae* lives on blades of Turtle Grass; and *Z. sunderlandi* lives under slabs of both living and dead coral, slightly offshore, on living coral reefs. To date, *Z. sunderlandi* has only been collected off Utila Is., Honduras.

Superfamily **Volutacea**
Family **Olividae**
Subfamily **Olivinae**
Genus *Agaronia* Gray, 1839

Agaronia hilli new species (Plate 12, Figures 1, 2)

Material Examined: HOLOTYPE - Length 46 mm, width 23 mm, trawled by commercial shrimpers from 20 m depth off Roatan Is., Honduras, 1985, USNM 859860; PARATYPE - length 45 mm, same locality and depth as holotype, Leonard Hill collection.

Description: Shell average size for genus, inflated, bulliform, with small, acuminate spire; aperture very wide and flaring; body whorl smooth, with silky texture; spire whorls almost completely covered by thick enamel deposit; anterior one-fourth of shell with thick, shiny, enamel deposit, sharply separated from silky-textured body whorl; columella thin, with several narrow, twisted plications; shell color olive-gray with speckles and small zig-zag of darker gray; spire enamel dark brownish-black; columella at posterior of aperture marked with large, blackish-brown patch; anterior enamel deposit dark brown on dorsum, fading to tan near columella; interior of aperture white, bordered with brown along inner edge of lip; columella pale tan and white; early whorls white.

Etymology: Named for Mr. Leonard Hill, Miami, Florida, who recognized the species as new and kindly donated the holotype.

Discussion: This, and the following species, are the first known living *Agaronia* species from the western Atlantic. *Agaronia hilli* is closest to the West African *A. hiatula* (Gmelin), but differs in being a more inflated shell with a wider aperture, and by having the spire covered by dark brown enamel. The new species is actually closest to the Costa Rican Pliocene fossil, *A. costaricensis* (Olsson), and is apparently the descendant. *Agaronia hilli* differs from its Pliocene ancestor by being a much more inflated shell. *Agaronia propatula* (Conrad) from the Panamic Province, is also closely related and also has the dark posterior columellar blotch. The eastern Pacific species, however has a lower spire and tends to be narrower and less inflated.

Agaronia leonardhilli new species (Plate 12, Figures 3, 4)

Material Examined: HOLOTYPE - Length 44 mm, width 16 mm, trawled by commercial shrimpers off Roatan Is., Honduras, 1985, USNM 859861.

Description: Shell narrow and very elongated; spire high and elevated; one-half of spire whorls covered by enamel; anterior enamel deposit two-toned in color, with posterior half being dark brown and anterior half being gray or white; aperture narrow; columella narrow, with twisted plications; shell color gray with numerous large, dark brown zig-zag, longitudinal flammules; on dorsum, zig-zag flammules only extend down posterior half of shell, leaving wide, clear gray band around anterior half; edge of shoulder along suture marked with conspicuous, large, intermittent, dark brown patches; spire enamel grayish-brown; interior of aperture brown.

Etymology: Named for Leonard Hill of Miami, Florida.

Discussion: To find two new, sympatric *Agaronia* species in one small area of the Caribbean is certainly an exciting discovery. It is interesting to note that there were two sympatric *Agaronia* species living along the Caribbean coast of Central America in the Pliocene, and that both gene pools survived into the Recent in the Caloosahatchian Secondary Relict Pocket of Honduras. The aforementioned *A. hilli* definitely is a descendant of the inflated, bulbous *A. costaricensis.* The more slender *A. leonardhilli* is descended from the slender Pliocene species, *A. mancinella* (Olsson).

Agaronia leonardhilli differs from its sympatric relative, *A. hilli,* by being a much more slender shell with a narrower aperture, by having a much more protracted, elevated spire, by being a more colorful shell with dark zig-zag markings and dark sutural patches, and by having a two-toned anterior enameled area. *Agaronia testacea* (Lamarck) from the Panamic Province, which coexists with *A. propatula,* is the living species that is closest to *A. leonardhilli.* The Caribbean species differs from its Pacific relative in being a more slender shell with a higher spire, by having darker enamel on the spire, and by having a narrower aperture. *Agaronia hilli* is the Caribbean analogue species of the Panamic *A. propatula,* while *A. leonardhilli* is the Caribbean analogue species of *A. testacea.*

Subfamily **Ancillinae**
Genus *Ancilla* Lamarck, 1799
Subgenus *Amalda* H. and A. Adams, 1847

Ancilla (Amalda) pacei new species (Plate 11, Figures 15, 16)

Material Examined: HOLOTYPE - Length 25 mm, width 12 mm, trawled by commercial shrimpers from 50 m depth off northern coast of Roatan Is., Honduras, 1979, USNM 859862; PARATYPE - 1 specimen same locality and depth as holotype, Robert Pace collection.

Description: Shell stocky, heavy, fusiform in shape; spire elevated; main section of body whorl with silky texture; suture bounded by wide band of enamel along shoulder region; spire whorls comprising two distinct zones, lower one with same silky texture of main body whorl, upper one with shiny enameled texture; fascicular band well-developed, surrounding anterior end of shell; shell color pale salmon-orange on main section of body whorl and lower zone of spire; enameled subsutural shoulder and fascicular bands bright orange, marked with evenly-spaced, dark orange-brown flammules; columella white, with one large, twisted plication; interior of aperture pale salmon-orange; protoconch pale salmon in color, rounded, dome-shaped; operculum large, oval, dark tan in color.

Etymology: Named for Robert Pace of Miami, Florida.

Discussion: *Ancilla pacei* is the northernmost ancilline olivid known from the Caribbean region. This new species somewhat resembles *A. tankervillei* (Swainson) from Venezuela, but differs in being a much smaller, stockier, and heavier shell, in having a salmon-orange base color, in having a much narrower shoulder enamel band, and in having shoulder and fascicular enamel bands of a bright orange color with orange-brown flammules. *Ancilla pacei* is the only known western Atlantic species of *Ancilla* to have dark-colored flammules on the enamel bands. *Ancilla aquaensis* (Olsson), pliocene fossil from the Gatun Formation of Costa Rica, is similar to *A. pacei,* both in size and shape, and is probably a direct ancestor.

Family Mitridae

Genus *Dibaphimitra* Cernohorsky, 1970

Dibaphimitra janetae new species (Plate 11, Figures 11, 12)

Material Examined: HOLOTYPE - Length 45 mm, width 19 mm, trawled by commercial shrimpers from 20 m depth off northern coast of Roatan Is., Honduras, 1984, USNM 859867; PARATYPES - 1 specimen, Roatan Is., Kremer collection; 1 specimen, Roatan Is., Robert Pace collection.

Description: Shell fusiform, almost biconic; spire elevated; whorls rounded, without shoulder; suture impressed, giving scalariform appearance to spire; body whorl and spire whorls with abundant minute spiral threads, giving shell a silky appearance; shell white, overlaid with scattered, large, amorphous blotches of deep wine-red color; base color overlaid with 13 evenly-spaced rows of small, dark brown dots; edge of suture bounded by row of conspicuous large, dark brown dashes; columella poorly developed, with 7 plications; only last 2 (posterior-most) are well-developed and readily visible; interior of aperture pure white; protoconch and early whorls pure white.

Etymology: Named for Janet Kremer, of Crystal Lake, Illinois, a great admirer of the *Mitridae.*

Discussion: *Dibaphimitra janetae* is the second known species of what was previously considered a monotypic genus. *Dibaphimitra florida* (Gould) (Plate 11, figures 13, 14), which is only found in northern Cuba and the Florida Keys, was thought to be the only living species. The new Roatan Island endemic species differs from *D. florida* in a number of consistent ways; by being a more slender shell without an angled shoulder; in having a white base color instead of violet or pink as in *D. florida;* in having a white protoconch instead of a violet or purple one; by having a characteristic row of large, dark brown dots and dashes along the margin of the suture; and having large wine-red patches on the shell. Although lacking the beautiful violet base color, protoconch, and aperture of *D. florida, D. janetae* is still a much more colorful shell, especially so with its large red patches and dotted suture margin. To date, *D. janetae* has only been collected of Roatan Is., where it is probably endemic.

Family **Volutidae**

Subfamily **Volutinae**

Genus *Falsilyria* Pilsbry and Olsson, 1954

Falsilyria harasewychi new species (Plate 15, Figures 1, 2)

Material Examined: HOLOTYPE - Length 45 mm, trawled by commercial shrimpers from 60 m depth off north coast of Roatan Is., Honduras, 1979, USNM 784485; PARATYPES - length 42 mm, same locality and depth as holotype, collection of Dr. Emilio F. Garcia; 1 specimen, Leonard Hill collection; 1 specimen Robert Pace collection.

Description: Shell small for subgenus, elongated, thickened; body whorl shiny, polished, with few thin, low, axial ribs; shoulder with 8-10 low, rounded knobs; subsutural area with 2-3 large, spiral cords; columella curving at posterior end, with 10-12 thin, heavily pustulated plications; shell white, overlaid with 8-10 evenly-spaced bands of large black and white dashes; spaces between bands of dashes completely covered with tiny, dark brown speckles; speckled and dashed color pattern, in turn, overlaid with interrupted bands and flammules of bright red; columella and aperture white; edge of lip with numerous, evenly-spaced black dashes; protoconch large, salmon-orange, dome-shaped.

Etymology: Named for Dr. M. G. Harasewych, Division of Mollusks, NMNH, Smithsonian

Institution.

Discussion: *Falsilyria harasewychi* is closest to *F. morrisoni,* but differs in being a consistently smaller and stockier shell with a much lower spire, and by having a sharper, more well-defined banded color pattern that lacks the large brown patches seen on *F. morrisoni.* The columella of *F. harasewychi* is curved at the posterior end, while that of *F. morrisoni* is only slightly curved. *Falsilyria morrisoni* is also a much more heavily sculptured shell, with numerous, close-packed, thin bladelike axial ribs, while *F. harasewychi* is a much smoother shell with only a few, or no, axial ribs. The shoulder knobs of *F. morrisoni* are also larger and more pointed than those of F. harasewychi. Another good distinguishing character is seen in the form of the lip; in *F. morrisoni,* the lip is more flaring, and extends beyond the posterior canal; in *F. harasewychi,* the lip is not flaring and is adherent to the side of the body whorl.

Due to the lack of specimens for study, I originally considered this small species to be simply a dwarf specimen of *F. morrisoni,* and made the holotype of this new species a paratype of *F. morrisoni.* Since that description, I have seen several more *F. harasewychi* in private collections, all nearly identical in size, color, and form, and have decided that they represent a valid new species. Interestingly enough, *F. harasewychi* appears more often in private collections than does *F. morrisoni.*

Falsilyria hilli new species (Plate 12, Figures 5, 6; Plate 14, Figure 9)

Material Examined: HOLOTYPE - Length 88 mm, width 42 mm, trawled by commercial shrimpers from 35 m depth of Punta Patuca, Honduras, 1985, USNM 859871; PARATYPES - 2 specimens, Strong collection; 1 specimen Leonard Hill collection.

Description: Shell oval in shape, inflated, shiny, with high polish; body whorl ornamented with numerous low axial plications; shoulder with 10-12 large, rounded knobs; spire whorls heavily knobbed; margin of suture bordered with 4 large, raised, spiral cords; columella with 10-12 large plications, often with smaller, secondary plications between; shell base color salmon-pink overlaid with two, wide, bright orange bands, one around mid-body, one around anterior end; wide orange bands overlaid with numerous thin, pale brown spiral lines;base of each shoulder knob with conspicuous, large, dark brown blotch; areas between wide orange bands with abundant tiny brown fleckles and rows of alternating dark brown and white dashes; columella and aperture bright salmon-pink; edge of lip marked with row of large dark-brown or black checkers; protoconch cylindrical in shape with flattened top, yellow-salmon in color.

Etymology: Named for Mr. Leonard Hill of Miami, Florida, who kindly donated the holotype.

Discussion: *Falsilyria hilli* is a new member of the beautiful and close-knit species complex comprising *F. demarcoi* (Olsson) (Plate 13, Figures 5, 6), *F. retemirabilis* Petuch (Plate 14, Figures 7, 8), *F. kotorai* (Petuch) (Plate 14, Figures 5, 6), *F. morrisoni* Petuch (Plate 15, Figures 3, 4), and *F. harasewychi* Petuch (Plate 15, Figures 1, 2) (described in this book). *Falsilyria hilli* is closest to *F. demarcoi,* but differs in being a pale salmon-pink shell instead of bright orange, and by consistently having the two dark orange, wide bands. The shoulder knobs of *F. hilli* are also larger and better developed than those of *F. demarcoi. Falsilyria hilli* is also similar to the following new species, *F. sunderlandi* (Plate 14, Figures 1, 2), but differs in being a much larger,more elongated shell, and by having the two wide dark orange bands.

Falsilyria sunderlandi new species (Plate 14, Figures 1, 2)

Material Examined: HOLOTYPE - Length 53 mm, on sand bottom, by SCUBA, 10 m depth off southern coast of Utila Island, Bay Islands, Honduras, 1986, USNM 859903; PARATYPES - 10 specimens, same depth and locality as holotype, Kevan Sunderland collection.

Description: Shell small for genus, short and stocky; whorls with 9 large, rounded axial ribs per whorl; axial ribs angled at shoulder, knoblike; subsutural area between shoulder and suture with numerous large, raised spiral cords; anterior end of shell with 6 incised sulci; rest of body whorl smooth, without spiral sculpture; base color white or pale pink with bands of large, evenly-spaced brown dots that are, in turn, separated by 3 evenly-spaced, solid, reddish-brown lines around mid-body; solid mid-body lines connected together by evenly-spaced, large, pale purple spots; columella with 10 large, smooth, plications, with several secondary plications in between; interior of aperture and columella white; protoconch large, rounded, dome-like, white in color.

Etymology: Named for Kevan Sunderland of Ft. Lauderdale, Florida, who collected the holotype on Utila.

Discussion: Of the eight known species of living *Falsilyria, F. sunderlandi* is the stockiest and most rotund. It is closest to *F. demarcoi* from Roatan Island (Plate 13, Figures 5, 6), but differs in being a much smaller, more obese shell with stronger axial ribs and shoulder knobs, in having a lower spire, in having fewer and coarser columellar plications, in having a narrower and less-developed mid-body color band, in having a white or pinkish-white base color instead of bright orange, and in having a white protoconch instead of an orange one. Although the islands of Utila and Roatan are only about 35 km apart, they are separated by deep water. This deep water has apparently acted as barrier, causing genetic isolation between *F. demarcoi* and *F. sunderlandi. Falsilyria sunderlandi* is endemic to Utila Island.

Genus *Voluta* Linnaeus, 1758

Voluta lindae new species (Plate 13, Figures 7, 8)

Material Examined: HOLOTYPE - Length 46 mm, width 24 mm, trawled by commercial shrimpers from 35 m depth off Bocas del Toro, Panama, 1976, USNM 859813; PARATYPE - fragment, shoulder and portion of spire, trawled by *R/V John Elliot Pillsbury,* from 50 m depth of Colón, Panama, 1963, invertebrate museum collection (wet collection), RSMAS, University of Miami.

Description: Shell small for genus, stocky with wide, sharp-angled shoulder; shoulder ornamented with 10 large, point knobs; body whorl below shoulder (anterior to) heavily sculptured with large, well-developed, axial plications and numerous large spiral cords; axial plications and spiral cords intersect to produce reticulated, fenestrate sculpture pattern; outer lip flaring; shell color bright canary yellow with band of large tan checkers just below shoulder and numerous, pale tan patches scattered over body whorl; bright yellow and tan base color overlaid with 8 evenly-spaced, thin bands of dark brown and yellow dashes; areas between dashed bands covered with dense masses of pale tan dots; area between shoulder and suture characteristically smooth, copiously covered with evenly-spaced tan dots; protoconch very large in proportion to shell size, protracted, cylindrical in shape; protoconch bright yellow; columella and aperture yellow; columella with 11 plications.

Etymology: Named for my wife, Linda J. Petuch.

Discussion: Based on the form of the protoconch and the reduced columellar plications, this new species appears to be referable to the genus *Voluta* and not *Falsilyria. Voluta lindae* is closest to *V. virescens* (Lightfoot) from southeastern Panama and northern Colombia (Plate 13, Figures 3, 4), but differs in being a smaller, stockier shell with a much larger knoblike protoconch, by having a smooth shoulder area, and by having a heavily reticulated sculpture pattern with strong, closely-packed axial riblets. *Voluta lindae* also differs in shell color, being bright yellow as opposed to lime green or gray-green as seen in *V. virescens.* The protoconch of *V. virescens* is small and dark brown, while that of *V. lindae* is large and bright yellow.

Voluta lindae is presently known only from off the coast of the Panama-Costa Rican border (Bocas del Toro area), but may extend along the Costa Rican and Nicaraguan coastlines in offshore areas. *Voluta lindae* is very similar to *V. alfaroi* Dall, a Pliocene fossil species from Costa Rica and Panama.

Family **Conidae**
Genus *Conus* Linnaeus, 1758

Conus kirkandersi new species (Plate 9, Figures 13, 14)

Material Examined: HOLOTYPE - Length 15 mm, width 9 mm, on exposed reef in tide pool, north end of Cozumel Is., Quintana Roo, Mexico, 1978, USNM 859873; PARATYPES - length 12, 14, and 15 mm, on beach, north end of Cozumel Is., collection of Danker Vink, Netherlands; 2 specimens, on beach, Cozumel Is., Robert Pace collection.

Description: Shell small for genus, thick, heavy, squat; spire low, almost flattened; shoulder rounded smooth, without coronations; body whorl sculptured with 12-15 large, raised spiral cords and numerous fine spiral threads, giving shell rough appearance; lip of adults thickened; shell color pure white with 2 broad bands, one above mid-body and one below mid-body; bands often break up into large brown blotches and flammules; anterior tip of shell brown; several rows of dark brown spots often superimposed upon brown bands; spire white with evenly-spaced, dark brown, crescent-shaped flammules; protoconch and early whorls yellow; interior of aperture white with 2 brown bands, corresponding to exterior color bands; periostracum thick, brown, tufted along shoulder; animal bright red.

Etymology: Named in memory of the late Kirk Anders of Ft. Lauderdale, Florida, who collected the holotype.

Discussion: *Conus kirkandersi* is one of the smallest members of a species-rich group of shallow-water, coral reef-associated cones that comprise the *C. magellanicus* Hwass species complex. This group of cones is characterized by a high degree of endemism, with most species being restricted to single islands or groups of islands throughout the Caribbean region. A biogeographical pattern of speciation such as this is most probably a function of low vagility coupled with geographical isolation due to glacial-eustatic sea level fluctuations during the late Pleistocene and early Holocene. As sea levels rose following a glacial stage, a shallow water ancestral species with direct development would have become isolated on island groups, and these populations would have diverged into sibling species. A similar biogeographic-parapatric speciation pattern was shown for the Central American *Falsilyria* volute species complex (PETUCH, 1981).

Of the member of the *C. magellanicus* species complex, *C. kirkandersi* is closest to *C. sphacelatus* Sowerby, a long lost species that has recently been collected on Grand Cayman Is., where it appears to be endemic (Plate 9, Figure 6). Being white with two rows of large, squarish, black checkers, *C. sphacelatus* closely resembles a small *C. ebraeus* Linnaeus from the Indo-Pacific and Panamic regions. Unlike *C. kirkandersi,* which has a smooth, noncoronated spire and sculpturing composed of raised cords, *C. sphacelatus* has a distinctly coronated spire and sculpturing composed of rows of tiny beads. The chocolate brown bands of *C. kirkandersi* also differ greatly from the black checkers of *C. sphacelatus. Conus kirkandersi* is also similar to *C. magnottei* Petuch, but the new Cozumel endemic differs from that species in being a rough-textured, corded shellas opposed to a smooth shell. The shell colors and color patterns also differ greatly, and these will be discussed separately in the description of *C. magnottei.* Like all of the small cones in the *C. magellanicus* complex, the animals of *C. kirkandersi* and *C. sphacelatus* are bright red.

Conus eversoni new species (Plate 10, Figures 12, 13, 14)

Material Examined: HOLOTYPE - Length 18 mm, collected by SCUBA diver among live *Agaricia* corals, 20 m depth at night, off south coast of Utila Is., Bay Islands, Honduras, 1986, USNM 859878; PARATYPES - 2 specimens, collected live under *Agaricia* coral, 20 m depth, off Utila Is., Honduras, Kevan Sunderland collection; 1 specimen same locality and depth as holotype, Wayne Harland collection; 1 specimen same locality and depth as holotype, Gene Everson collection.

Description: Shell small, slender, tapered, with sharp-angled shoulder; spire flattened with elevated, mamillate protoconch; body whorl smooth, polished, with 10 spiral cords around anterior end; spire whorls with 4 spiral cords; shell color dark reddish-brown with variable number of spiral rows of dark brown, tiny dots; paler reddish-brown band around mid-body; spire whorls with numerous evenly-spaced, dark brown flammules; early whorls and protoconch pale tan; interior of aperture purple.

Etymology: Named for Mr. Gene Everson, Matthews, North Carolina, who collected many new species in Honduras and Cocos Island.

Discussion: Although the holotype was dead and occupied by a hermit crab when collected, live *Conus eversoni* were collected deep within the living coral reef, usually hanging from the under sides of giant sheets of *Agaricia* (Lettuce Coral). *Conus eversoni,* then, is one of the few known Caribbean cones shells that actually lives among living corals in interstitial areas on coral reefs. *Conus eversoni* is closest to *Conus flamingo* Petuch in shape and size, but differs in being a brown-colored shell with rows of tiny dots, in having a purple aperture, and in having sculptured spire whorls with spiral cords. The ecologies of the two species are very different, with *Conus flamingo* living on sand bottoms in about 80 m depth and *C. eversoni* living deep within the reef amongst living corals, in about 20 m depth. At present, *C. eversoni* has only been collected on coral reefs off Utila and Guanaja Islands (Pace collection), Bay Islands, Honduras.

Conus harlandi new species (Plate 12, Figures 9, 10, 11, 12)

Material Examined: HOLOTYPE - Length 33 mm, width 17 mm, in sand near weed beds, 60 ft depth off Utila Is., Bay Islands, Honduras, 1986, USNM 859883; PARATYPES - 2 specimens, lengths 11, 26 mm, same locality and depth as holotype; Kremer collection; 10 specimens, same locality as holotype, Harland collection; 20 specimens, same locality as holotype, Sunderland collection, 2 specimens, same locality, USNM 859907.

Description: Shell tapered, elongated; spire low, almost flattened; edge of shoulder sharp, smooth, without coronations; body whorl polished, shiny, covered with numerous very fine, closely-packed spiral threads, giving shell silky appearance; shell color reddish-brown to chestnut, overlaid with 12-15 evenly-spaced bands of tiny, white flammules and dots; one wide white band around shoulder, one around anterior tip, each, in turn, containing tiny reddish-brown dots and hairlike flammules; some specimens (as in holotype) with white axial bands that correspond to previous lip edges; axial bands intersect spiral bands of white dots to produce checkered appearance; aperture long, narrow; interior of aperture with rose-pink shading; spire white, with scattered, large reddish-brown blotches and numerous thin, hairlike flammules in between; periostracum thin, transparent.

Etymology: Named for Wayne Harland, Lighthouse Point, Florida, who collected the holotype on Utila.

Discussion: *Conus harlandi* is the northernmost member of the predominantly southern Caribbean *C. cedonulli* Linnaeus species complex. This new Honduran species is closest in shell form to *C. sanctaemarthae* Vink from northern Colombia, and shares the same tapered body and low, noncoronated shoulder. *Conus sanctaemarthae,* however, is a beautiful lilac-purple shell, covered with numerous rows of brown and white dots, while *C. harlandi* is a reddish-brown shell with evenly-spaced bands of white flammules. The basic color pattern of *C. harlandi* resembles that of the much larger *C. cedonulli* from St. Vincent Is., but differs in being much finer and more regular.

Conus magnottei new species (Plate 12, Figures 7, 8)

Material Examined: HOLOTYPE - Length 14 mm, width 7 mm, 2 m depth under rocks, north coast of Roatan Is., Honduras, 1979, USNM 859887; PARATYPES - 1 specimen same locality and depth as holotype, Robert Pace collection; 1 specimen, same locality and depth as holotype, Kremer collection.

Description: Shell small for genus, squat in form; spire low, smooth, without coronations; body whorl smooth, shiny, without sculpturing; anterior tip with few small, low cords; color bright purplish-pink to lilac, with numerous white patches and flammules; white or pale pink band around mid-body; holotype with band of dark brown and white patches around mid-body; spire white, with some specimens having scattered dark brown flammules (such as holotype); protoconch and early whorls bright pink; interior of aperture purple.

Etymology: Named for Gary Magnotte of Pompano Beach, Florida, who collected the holotype on Roatan.

Discussion: *Conus magnottei* is closest to *C. kirkandersi* from Cozumel Is., but differs in lacking the corded sculpturing of *C. kirkandersi* and by having a bright pink or lilac color pattern instead of white and brown-banded one. *Conus magnottei* is endemic to Roatan Is.

Conus sunderlandi new species (Plate 12, Figures 13, 14, 15, 16)

Material Examined: HOLOTYPE - Length 33 mm, width 17 mm, in sand near weed beds, 60 ft depth off Utila Is., Bay Islands, Honduras, 1986, USNM 859892; PARATYPES - 5 specimens, same locality and depth as holotype, Wayne Harland collection; 12 specimens, same locality and depth as holotype, Sunderland collection.

Description: Shell stocky, solid, broad across shoulder; shoulder sharp-angled; body whorl smooth and shiny; anterior tip with several strong spiral cords; shell color white, overlaid with numerous close-packed rows of tiny, bright orange, vertical flammules; mid-body with white band; anterior tip white, with few scattered tiny, orange flammules; bands of orange vertical flammules sometimes coalesce to form solid orange, wide band (as in paratype shown here, Figures 15, 16); spire whorls white, with numerous, closely-packed, bright orange, crescent-shaped flammules; interior of aperture pink; periostracum thin, smooth, yellow.

Etymology: Named for Kevan Sunderland of Ft. Lauderdale, Florida, who collected the holotype.

Discussion: *Conus sunderlandi,* which appears to be endemic to Utila Is. and surrounding areas of the Bay Islands, is closest to *C. daucus,* Hwass, which is widespread throughout the Caribbean. *Conus sunderlandi* differs from *C. daucus* in being a slightly more pyriform shell with a higher spire, by being a smaller species, and by having a distinctive color pattern of bands of orange, vertical flammules. No known varieties of *C. daucus,* from anywhere in the Caribbean, show the banded pattern of *C. sunderlandi.* The color pattern of *C. sunderlandi* somewhat resembles that of the unrelated *C. lorenzianus* Dillwyn, which ranges from Honduras and Belize south to Venezuela. In *C. lorenzianus,* the bands of flammules are much coarser and larger, approximating the spotted pattern of *C. spurius* Gmelin.

Family **Turridae**
Subfamily **Clavinae**
Genus *Hindsiclava* Hertlein and Strong, 1955

Hindsiclava tippetti new species (Plate 13, Figures 12, 13)

Material Examined: HOLOTYPE - Length 70 mm, trawled from 35 m depth off Rosalind Bank, Honduras, by commercial shrimp trawler, 1978, USNM 784647; PARATYPE - 1 specimen, 55 m length, same location and depth as holotype, USNM.

Description: Shell large for genus, elongate, with elevated spire; shell ornamented with numerous equal-sized axial and spiral cords that are arranged in cancellate pattern; intersection of axial and spiral cord producing large, rounded bead; early whorls with large, beaded, axial ribs; axial ribs are replaced with cancellate-beaded sculpture on later whorls; subsutural area with large, wide, raised band and channeled, unsculptured area that corresponds to anal notch; anal notch wide, deep; aperture narrow; shell color dark brown with tan beads; early whorls and protoconch dark brown; interior of aperture brown.

Etymology: Named for Dr. Donn Tippett, Research Associate of the Division of Mollusks, Smithsonian Institution, and an authority on the *Turridae.*

Discussion: *Hindsiclava tippetti* is the largest and most colorful species of *Hindsiclava* known from the western Atlantic. It is closest to the widespread, deep water *H. alesidota* (Dall), but differs in being a much larger and wider species, in having a coarser, more ornate beaded sculpture pattern, by having a wider subsutural channel, and in having a dark brown and tan-beaded color pattern. *Hindsiclava tippetti* has been collected primarily on the Rosalind Banks off the Honduras-Nicaragua border, but may also be found on the surrounding offshore banks and the mainland of Central America.

PLATE 11. NEW GASTROPODS FROM THE BAY OF HONDURAS AND ROATAN ISLAND

Figures 1, 2: *Bolma (Senobolma) sunderlandi* Petuch, n.sp., dorsal and ventral views of holotype, 16 mm.

Figures 3, 4: *Distorsio robinsoni* Petuch, n.sp., dorsal and ventral views of holotype, 53 mm.

Figures 5, 6: *Distorsio clathrata* (Lamarck, 1816), dorsal and ventral views of 55 mm specimen, Big Pine Key, Florida Keys (for comparison with *Distorsio robinsoni).*

Figures 7, 8: *Murex samui* Petuch, n.sp., dorsal and ventral views of holotype, 76 mm.

Figures 9, 10: *Murex garciai* Petuch, n.sp., dorsal and ventral views of holotype, 45 mm.

Figures 11, 12: *Dibaphimitra janetae* Petuch, n.sp., dorsal and ventral views of holotype, 45 mm.

Figures 13, 14: *Dibaphimitra florida* (Gould, 1856), dorsal and ventral views of 41 mm specimen, Pickles Reef, off Tavernier Key, Florida Keys (for comparison with *Dibaphimitra janetae).*

Figures 15, 16: *Ancilla (Amalda) pacei* Petuch, n.sp., dorsal and ventral views of holotype, 25 mm.

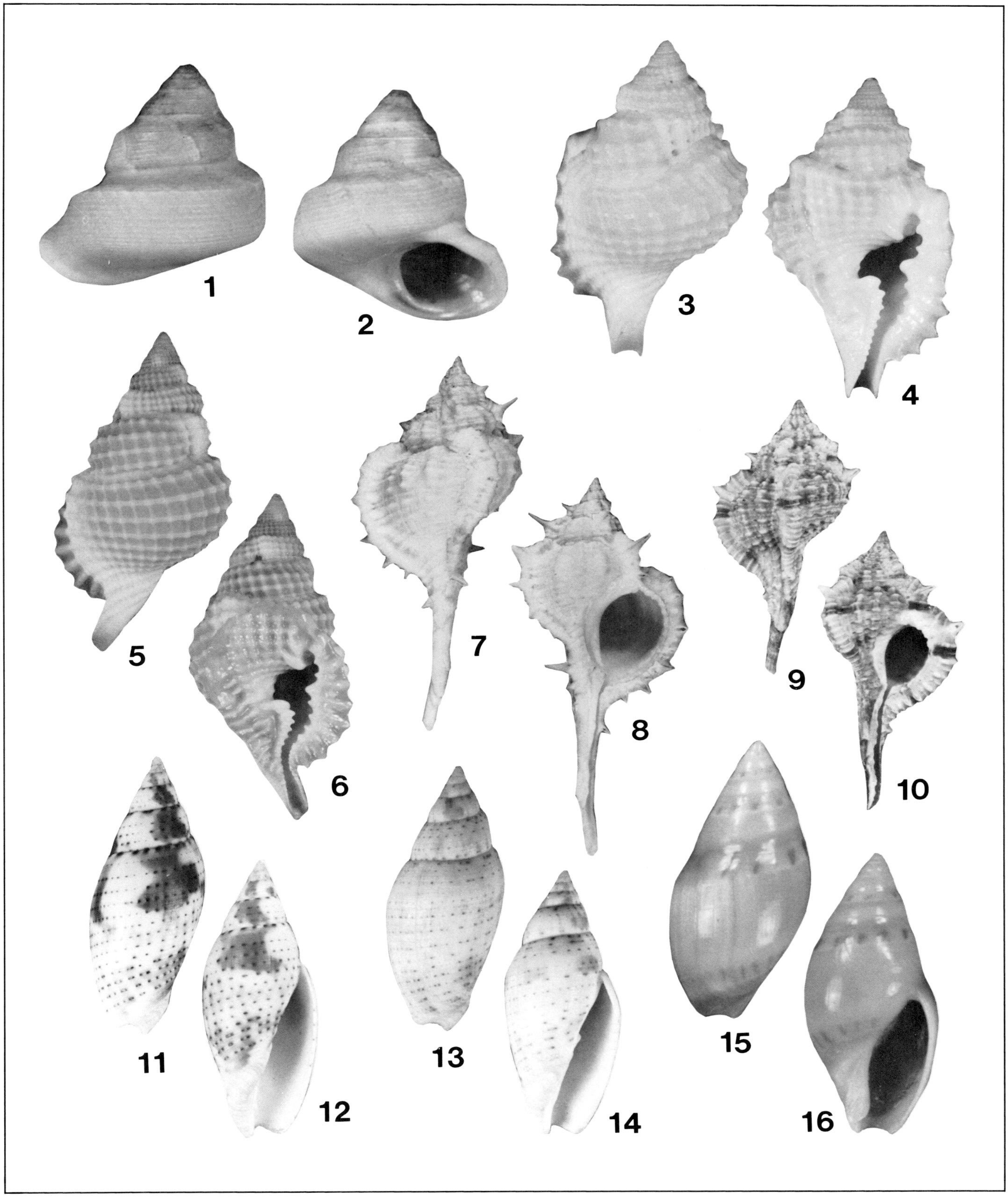
1
2
3
4
5
6
7
8
9
10
11
12
13
14
15
16

PLATE 12. NEW GASTROPODS FROM THE BAY OF HONDURAS AND ROATAN ISLAND

Figures 1, 2: *Agaronia hilli* Petuch, n.sp., dorsal and ventral views of holotype, 46 mm.

Figures 3, 4: *Agaronia leonardhilli* Petuch, n.sp., dorsal and ventral views of holotype, 44 mm.

Figures 5, 6: *Falsilyria hilli* Petuch, n.sp., dorsal and ventral views of holotype, 88 mm.

Figures 7, 8: *Conus magnottei* Petuch, n.sp., dorsal and ventral views of holotype, 14 mm.

Figures 9, 10: *Conus harlandi* Petuch, n.sp., dorsal and ventral view of paratype, 30 mm.

Figures 11, 12: *Conus harlandi* Petuch, n.sp., dorsal and ventral views of holotype, 33 m.

Figures 13, 14: *Conus sunderlandi* Petuch, n.sp., dorsal and ventral views of holotype, 33 mm.

Figures 15, 16: *Conus sunderlandi* Petuch, n.sp., dorsal and ventral views of dark colored paratype.

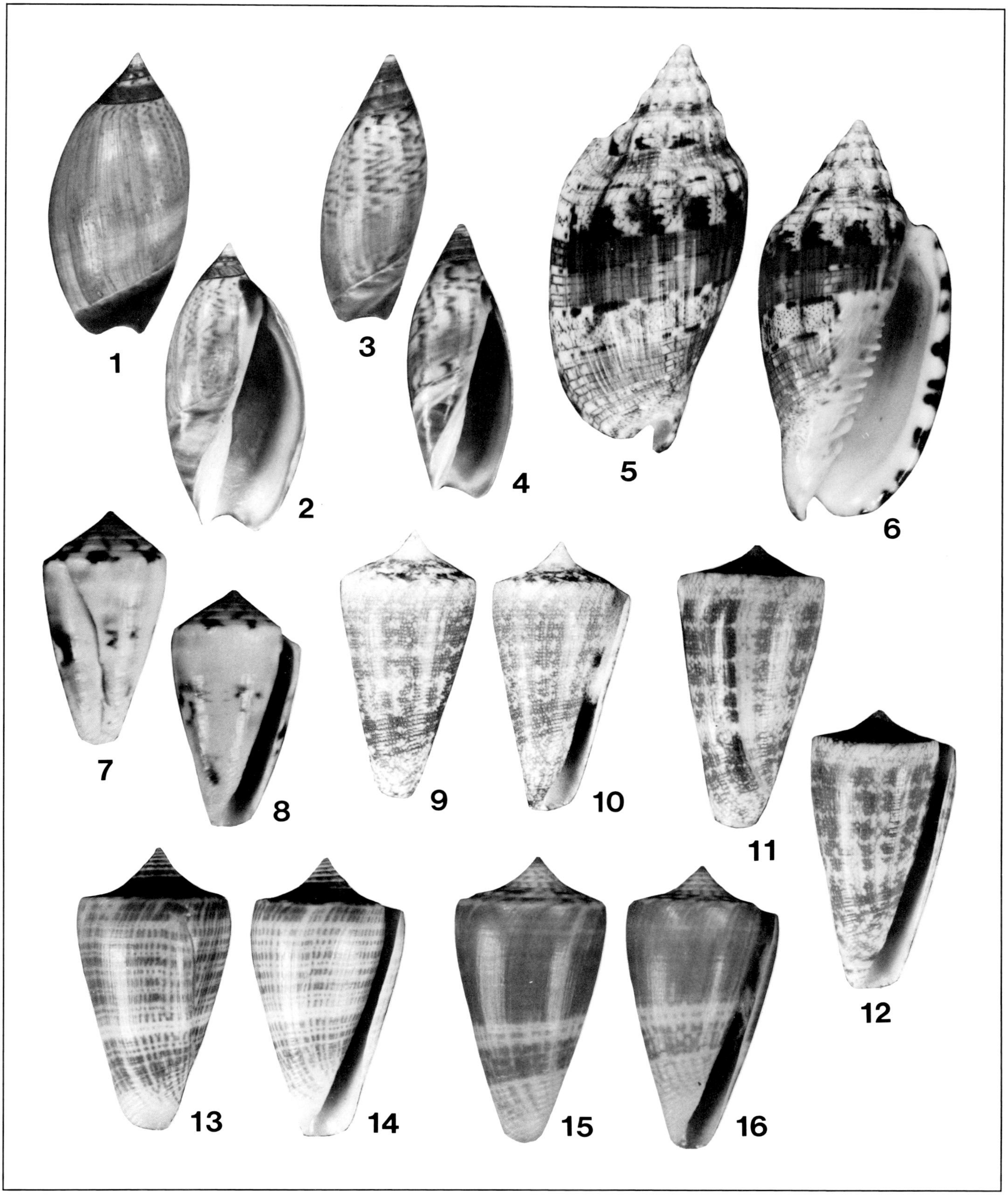
1
2
3
4
5
6
7
8
9
10
11
12
13
14
15
16

PLATE 13. NEW GASTROPODS FROM EASTERN CENTRAL AMERICA AND NORTHERN SOUTH AMERICA

Figures 1, 2: *Chicoreus emilyae* Petuch, n.sp., dorsal and ventral views of holotype, 55 mm.

Figures 3, 4: *Voluta virescens* Lightfoot, 1786, dorsal and ventral views of 57 mm specimen, Cartagena, Colombia.

Figures 5, 6: *Falsilyria demarcoi* (Olsson, 1965), dorsal and ventral views of 85 mm specimen, Roatan Is., Honduras.

Figures 7, 8: *Voluta lindae* Petuch, n.sp., dorsal and ventral views of holotype, 46 mm.

Figures 9, 10: *Harasewychia harasewychi* Petuch, n.gen. and n.sp., dorsal and ventral view of holotype, 34 mm.

Figure 11: *Aphera lindae* Petuch, n.sp., ventral view of holotype, 10 mm.

Figures 12, 13: *Hindsiclava tippetti* Petuch, n.sp., dorsal and ventral views of holotype, 70 mm.

1
2
3
4
5
6
7
8
9
10
11
12
13

PLATE 14. VOLUTE SHELLS FROM EASTERN CENTRAL AMERICA

Figures 1, 2: *Falsilyria sunderlandi* Petuch, n.sp., dorsal and ventral views of holotype, 53 mm, Utila Is, Honduras.

Figures 3, 4: *Falsilyria kotorai* Petuch, 1981, dorsal and ventral views of holotype, 49 mm, Rosalind Bank, Honduras.

Figures 5, 6: *Falsilyria garciai* Petuch, 1981, dorsal and ventral views of holotype, 71 mm, Gorda Bank, Honduras.

Figures 7, 8: *Falsilyria retemirabilis* Petuch, 1981, dorsal and ventral views of holotype, 75 mm, Caratasca Cays, Honduras.

Figure 9: *Falsilyria hilli* Petuch, n.sp., dorsal view of 85 mm specimen, Punta Patuca, Honduras.

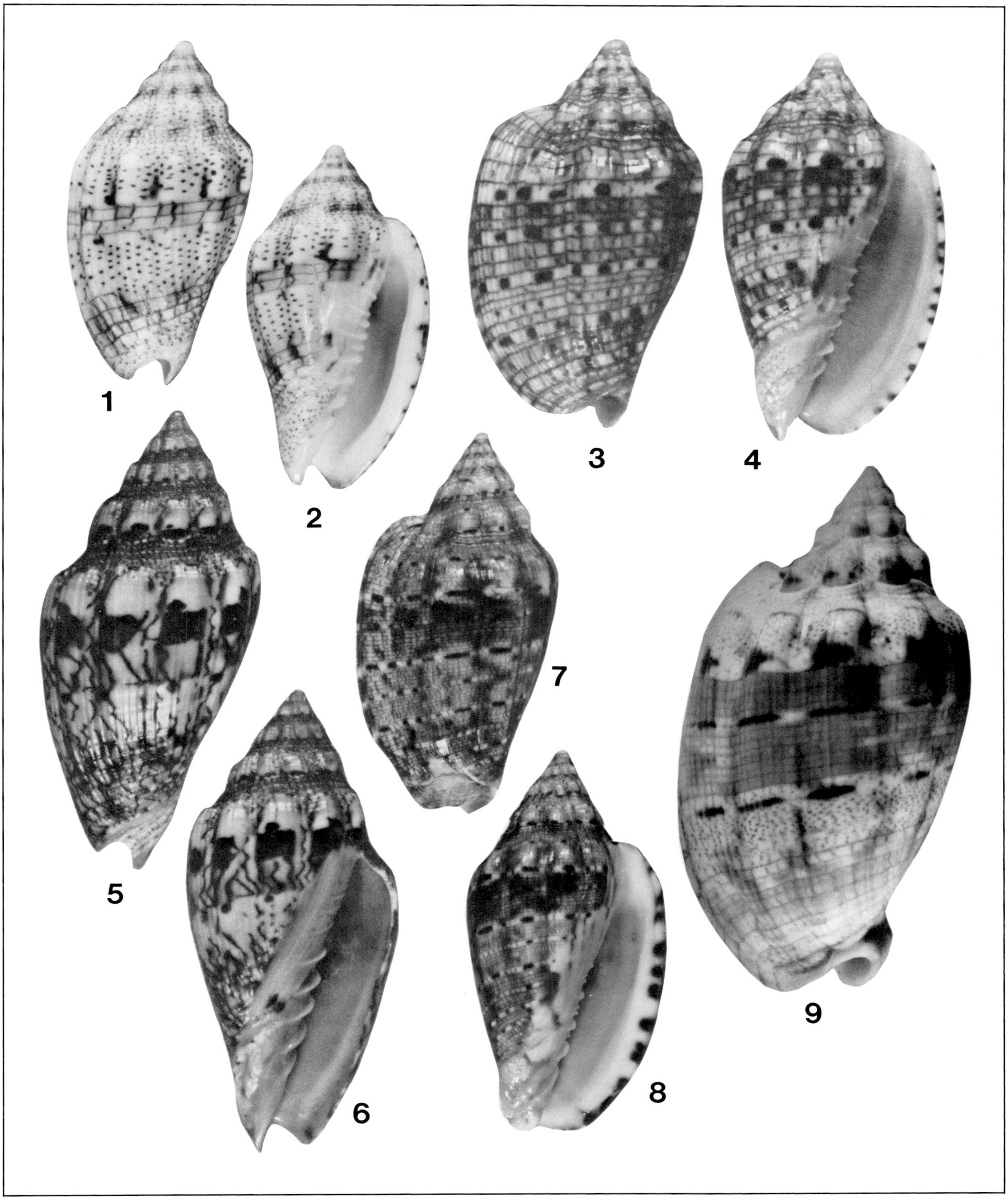

PLATE 15. NEW AND SELDOM-SEEN GASTROPODS FROM HONDURAS, BARBADOS AND SOUTH AMERICA

Figures 1, 2: *Falsilyria harasewychi* Petuch, n.sp., dorsal and ventral views of holotype, 45 mm.

Figures 3, 4: *Falsilyria morrisoni* Petuch, 1981, holotype, 74 mm, 60 m depth off Roatan Is., Honduras.

Figures 5, 6: *Lindapterys sanderi* Petuch, n.gen. and n.sp., dorsal and ventral views of holotype, 11mm.

Figures 7, 8: *Sconsia lindae* Petuch, n.sp., dorsal and lateral views of holotype, 71 mm.

Figure 9: *Panamurex gatunensis* (Brown and Pilsbry, 1911), ventral view of 26 mm specimen, 35 m depth, Golfo de Triste, Venezuela.

Figures 10, 11: *Lindapterys vokesae* Petuch, n. gen. and n.sp., dorsal and ventral views of holotype, 11.7 mm, fossil, Chipola Formation of Florida (type of genus).

Figures 12, 13: *Conus kulkulcan* Petuch, 1980, dorsal and ventral views of holotype, 21 mm, Roatan Island, Honduras.

Figure 14: *Conus kulkulcan* Petuch, 1980, ventral view of paratype, 22 mm, Roatan Is., Honduras.

Figures 15, 16: *Muricopsis oxossi* Petuch, 1979, dorsal and ventral views of holotype, 9 mm, Abrolhos Archipelago, Brazil.

Figures 17, 18: *Cyphoma macumba* Petuch, 1979, dorsal and ventral views of holotype, 22 mm, Abrolhos Archipelago, Brazil.

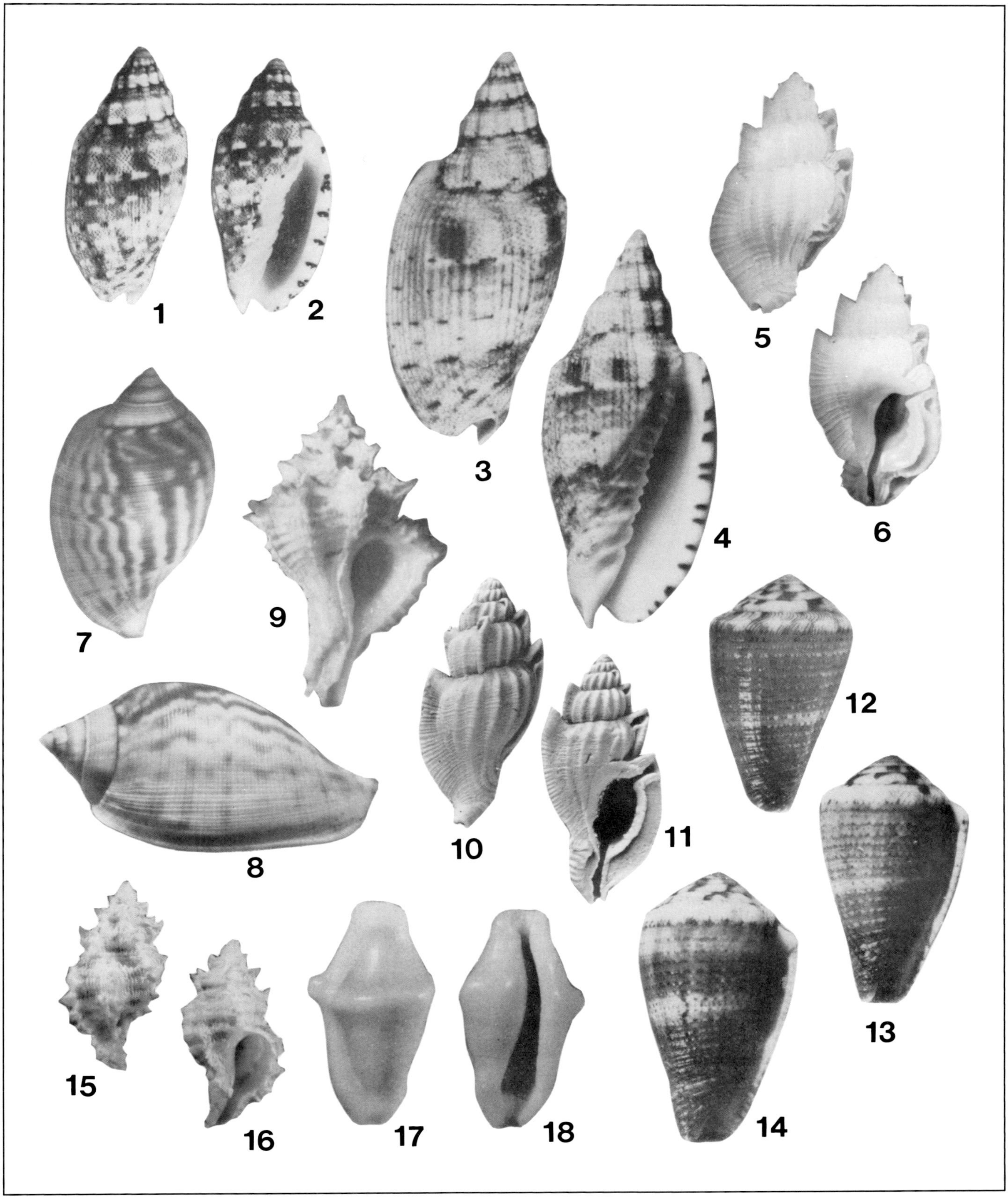
1
2
3
4
5
6
7
8
9
10
11
12
13
14
15
16
17
18

CHAPTER 6

New Discoveries from Northern South America

The northern coast of South America, from the Panama-Colombia border at the Gulf of Uraba to the Orinoco River, Venezuela, probably contains the least-known molluscan fauna in the Caribbean region. Most of this region belongs to the Gatunian Relict Pocket (discussed in Chapter 2) and contains many species groups that were thought to have been extinct in the Caribbean region since the Pliocene. The discovery of these prehistoric, living fossils along northern South America led to my proposing the concept of geographical heterochrony in marine zoogeography (PETUCH, 1982).

The Colombian and Venezuelan coastlines comprise four distinct subregions that are mostly demarcated by substrate type and water temperature in relation to the upwelling systems. Most of the material shown in this chapter was collected by commercial shrimpers as they drag their nets over large areas of the sea bottoms in these regions. The area along Colombia, from the Gulf of Uraba to Santa Marta, contains a muddy bottom with relatively warm waters that are not in the influence of upwellings. The muddy bottom is a result of the large sediment fan that emanates from the mouth of the Magdalena River, which empties into the Caribbean near Barranquilla. The muddy effluent of the Magdalena River is carried southward along the Colombian coast by currents.

Like the Goajira Peninsula, the Gulf of Venezuela also comes under the influence of upwelling systems. The sheltered, quiet bays along the western side of the Paraguana Peninsula contain cool, nutrient-rich water that is rich in plankton, producing murky diving conditions. For this reason, working on commercial shrimp boats in the Gulf of Venezuela is still the best way to sample the bottom fauna. The sea floor off the western side of the Paraguana is mostly clay-like, diatom-rich mud with large areas made up of *Turritella* and scallop beds and large forests of tubular sponges. In the shallow bays, such as Amuay Bay, Turtle Grass *(Thalassia testudinum)* forms lush beds that carpet the entire bottom.

The fourth subregion, the Gulf of Triste off Carabobo State, Venezuela, is similar in substrate and oceanographic conditions to the Gulf of Venezuela. The outer edge of the Gulf, however, contains one of the richest molluscan assemblages yet found within the Caribbean region. The shallow western side of the Gulf of Triste contains warmer, protected water and, especially near Tucacas, large Turtle Grass beds like those along the western side of the Paraguana Peninsula.

As the shrimpers ply the fertile waters of these four subregions, they encounter a number of characteristic, ubiquitous gastropods. These most often include the large spindle shell *Fusinus caboblanquensis* Weisbord, the spiny form of *Murex donmoorei* Bullis, the large cowrie *Cypraea (Siphocypraea) donmoorei* (Petuch), and the large, obese form of the olive shell, *Oliva tisiphona* form *oblonga* Marrat. These last are shown on Plates 18, 19, and 20. Along the muddy coast south of the Magdalena River mouth, the marginellids *Persicula porcellana* (Gmelin) (Plate 19, Figure 16) and *Marginella marginata* (Born) (Plate 24, Figures 15, 16) are commonly trawled, as are *Voluta virescens* Lightfoot (Plate 18, Figure 11; Plate 13, Figures 3, 4). The rare, giant volute, *Scaphella (Clenchina) evelina* Bayer (Plate 18, Figure 2; Plate 21, Figure 1), has also been taken along this coast. In the region of the Goajira Peninsula and Gulf of Venezuela, an entire endemic cone shell radiation is present. Along with the new species described in this book, a number of rarely-seen cone shells are frequently collected by shrimpers (but, unfortunately, usually quickly dumped back into the sea with other flotsam and jetsam brought up in the nets). These include *Conus undatus* Kiener (Plate 19, Figure 12), *Conus granarius* Kiener (Plate 20, Figures 8, 13), *Conus sanguineus* Kiener (a lost species) (Plate 20, Figure 10), a number of *Conus mappa* Lightfoot subspecies (Plate 20, Figure 14), *Conus gibsonsmithorum* Petuch (Plate 25, Figure 7), *Conus penchaszadehi* Petuch (Plate 16, Figures 14, 15), *Conus hedgesi* Sowerby (Plate 21, Figure 4), *Conus puncticulatus* Hwass (Plate 19, Figures 4, 5), and the dark form *Conus puncticulatus* form *mauritianus* Hwass (Plate 19, Figure 6), *Conus cingulatus* Lamarck, *Conus lorenzianus* Dillwyn, and *Conus centurio* Born.

Equally as impressive as the cone shell fauna from along this coast is the murex shell fauna. Frequently collected along with the cones in the same dredge haul are *Siratus thompsoni* Bullis (Plate 20, Figure 3), *Siratus springeri* Bullis (Plate 18, Figure 6), *Panamurex velero* E. Vokes, *Panamurex recticanalis* (Weisbord), *Siratus briskasii* (Verrill) (another lost species) (Plate 18, Figures 9, 10; Plate 21, Figure 13), *Phyllonotus cf. globosus* (Emmons) (possibly a new species) (Plate 19, Figures 1, 2), *Murex olssoni* E. Vokes, and *Murex blakeanus* E. Vokes. The buccinid genus *Antillophos* is particularly common on the soft bottom areas off Colombia and Venezuela. Most frequently encountered in the shrimp nets are *Antillophos candei* (Orbigny) (Plate 24, Figure 6), *Antillophos adelus* (Schwengel) (Plate 24, Figures 7, 8), and *Antillophos elegans* (Guppy) (Plate 24, Figures 9, 10). The auger shell *Terebra (Strioterebrum) petiveriana* Deshayes (Plate 20, Figure 11), the olivids *Eburna glabrata* (Linnaeus) (Plate 19, Figure 7), and *Amalda tankervillei* (Swainson) (Plate 19, Figure 9), and the turrids *Fusiturricula iole* Woodring (Plate 20, Figure 17), *Clathodrillia gibbosa* (Born) (Plate 20, Figure 9), *Hindsiclava chazaliei* (Dautzenberg) (Plate 19, Figure 13), and *Polystira cf. barretti* (Guppy) (Plate 19, Figure 14) are also common along this coast.

The Gulf of Venezuela and Gulf of Triste areas also contain a large number of species that are not found farther south along the Colombian coast of the Magdalena River area. Commonly collected in the Gulf of Venezuela by shrimp boats, and also along the northern side of the Goajira Peninsula, are the paciphilic columbellids *Strombina pumilio* (Reeve) (Plate 20, Figure 16), *Strombina caboblanquensis* Weisbord (Plate 20, Figure 15), and *Strombina galba* Weisbord (Plate 18, Figures 13, 14) (which was originally described from a juvenile), the unusual marginellids *Cryptospira martini* (Petit) (Plate 24, Figures 13, 14), and *Persicula multilineata* (Sowerby), the large blue-colored *Fasciolaria tulipa hollisteri* Weisbord (Plate 20, Figure 6), the volutaceans *Volutomitra erebus* Bayer (Plate 23, Figure 7), and *Voluta musica guinaica* Lamarck (bright blue, smooth-shouldered subspecies) (Plate 20, Figure 7), the bizarre turbinellid *Fulgurofusus brayi* (Clench) (Plate 18, Figures 2, 3, 4; Plate 21, Figures 2, 3, 7) (found in shallower water in upwelling areas), the Venezuela endemic auger shells *Terebra (Strioterebrum) trispiralis* Weisbord (Plate 20, Figure 18), and *Terebra (Strioterebrum) quadrispiralis* Weisbord (Plate 20, Figure 19), and the rare, bright red-orange olivid *Amalda venezuelana* (Weisbord) (Plate 19, Figure 8). The shallow water, Turtle Grass environment from Riohacha on the outer coast of the Goajira Peninsula around to the western coast of the Paraguana Peninsula are home to the beautiful *Cypraea (Siphocypraea) mus* Linnaeus (Plate 22, Figures 3, 4), which can often be seen crawling on exposed weed beds, at low tide, by the thousands. All of these gastropods combine to give a distinctive flavor to the molluscan fauna of Colombia and Venezuela, and set this area apart, biogeographically, from the rest of the Caribbean (see PETUCH, 1982).

Besides these known species, many other important biotic discoveries have also come to light recently, and these help to underscore the uniqueness of the fauna of this area. Some of these discoveries, which will be discussed later in this chapter under their individual descriptions, include: the first *Murexsul* species from the Caribbean region, the first member of the Arctic genus *Mohnia* to be collected within the Caribbean, the first Atlantic species of *Truncaria,* the first Atlantic species of the Indo-Malaysian olivid genus *Gracilancilla,* the first Atlantic species of the Japonic-Australian genus *Turrancilla,* the first living specimens of the supposedly extinct genera *Conomitra* and *Paraborsonia,* and the first Atlantic species of the supposedly-paciphilic genus *Aphera.*

An unusual bivalve fauna is also present along northern South America. One of the more unusual groups that frequently washes up on Colombian beaches is the nuculanid genus *Adrana,* which is usually a deeper water, offshore genus. At one beach near Cartagena (Boquilla), four species of *Adrana,* one of which was new to science and is described here, can be found together abundantly in windrows along the hightide line. Mixed in with these odd nuculanids, the gastropods *Voluta virescens* and *Conus cingulatus* are also commonly collected in these windrows, giving an indication of the strange ecosystem that must exist just offshore. The three commonest nuculanids encountered on the Cartagena beaches are *Adrana scaphoides* Rehder (Plate 26, Figure 5A), *Adrana notabilis* Rehder (Plate 26, Figure 4), and *Adrana tellinoides* (Sowerby) (Plate 26, Figure 5). Scallops are also abundant and characteristic of the shallow water areas along northern South America. The large *Amusium laurenti* (Gmelin), *Aequipecten lineolaris* (Lamarck), and *Pacipecten linki* (Dall), are all abundant along this coast, form large beds that often carpet the bottom, and are commercially fished by shrimpers during their off-season.

The Netherlands Antilles islands of Aruba, Curaçao, and Bonaire are included here as the insular component of the continental fauna. The island of Aruba, in particular, contains many rare or endemic species that appear to be more closely related to South American species than those from the West Indies. Along the Malmok coast of Aruba, the endemic Netherlands Antilles olivids *Eburna balteata* (Swainson) (Plate 17, Figure 1), *Oliva fulgurator* (Röding) and *Oliva reclusa* Marrat (Plate 17, Figure 4), are especially common, as is *Murexiella hilli,* a beautiful new muricid described here. These islands are in much need of scientific investigation, and I predict that many new and scientifically interesting species will be discovered there in future years.

The island of Barbados, although technically an isolated seamount, is included here because of its proximity to South America. The deep water areas around the island have recently yielded a number of amazing gastropods, including *Sassia lewisi* Harasewych and Petuch (Plate 16, Figures 1, 2, 3), *Sconsia nephele* Bayer (Plate 23, Figure 5), *Latirus bernadensis* Bullock, (Plate 18, Figure 1), *Oliva barbadensis* Petuch and Sargent, and *Conus mazei* Deshayes (Plate 7, Figure 7). Of special interest from Barbados was the discovery of a completely new muricid genus, here named *Lindapterys,* that previously was known only from fossils from Florida and Austria. Like the Netherlands Antilles islands of Aruba, Curaçao, and Bonaire, the deep water areas around Barbados are virtually unexplored and are sure to yield many new species in years to come.

New Species From Northern South America

Homalopoma finkli new species
Phenacovolva (Turbovula) lindae new species
Pseudocyphoma gibsonsmithorum new species
Cypraea (Siphocypraea) mus tristensis new subspecies

Morum (Cancellomorum) lindae new species
Sconsia lindae new species
Dermomurex (Trialatella) kaicherae new species
Lindapterys sanderi new genus and new species
Murex sunderlandi new species
Murexsul harasewychi new species
Harasewychia harasewychi new genus and new species
Colubraria bayeri new species
Antillophos bayeri new species
Mohnia kaicherae new species
Truncaria lindae new species
Gracilancilla lindae new species
Turrancilla williamsoni new species
Subcancilla leonardhilli new species
Subcancilla lindae new species
Nodicostellaria kremerae new species
Turricostellaria leonardhilli new species
Turricostellaria lindae new species
Conomitra leonardhilli new species
Conomitra lindae new species
Aphera lindae new species
Conus aureopunctatus new species
Conus bayeri new species
Conus colombianus new species
Conus finkli new species
Conus kevani new species
Conus paraguana new species
Conus parascalaris new species
Conus perprotractus new species
Conus tristensis new species
Conus venezuelanus new species
Paraborsonia lindae new species
Polystira lindae new species
Scissuladrana ludmillae new species

New Species From The Netherlands Antilles

Murexiella hilli new species
Muricopsis aliceae new species
Oliva sargenti new species

Family **Turbinidae**
Subfamily **Homalopomatinae**
Genus *Homalopoma* Carpenter, 1864

Homalopoma finkli new species (Plate 26, Figures 13, 14)

Material Examined: HOLOTYPE - Height 18 mm, width 19 mm, trawled from 50 m depth by commercial shrimpers, off Puerto Cabello, Golfo de Triste, Venezuela, 1978, USNM 859815.

Description: Shell very large for genus, wide, inflated, turbinate in shape; spire narrow in proportion to shell width, elevated; body whorl ornamented with 8 large, well-developed, raised cords; spire whorls ornamented with 3 cords; outer layer of shell rough-textured, light brown in color; inner layer of shell nacreous, with nacre showing through thin areas on spire and early whorls; outer lip and aperture flaring; columella thickened, nacreous, partially covered with white shelly layer; interior of aperture nacreous, iridescent; columella with one small tooth.

Etymology: Named for Dr. Charles Finkl, Jr., of the Coastal Education and Research Foundation (CERF), Charlottesville, Virginia.

Discussion: *Homalopoma finkli* is by far, the largest known Atlantic *Homalopoma,* being at least four to five times larger than any known species. It is also found in much shallower water than is normal for this outer continental shelf and slope group. *Homalopoma finkli* is closest to widespread *H. linnei* (Dall), but differs in being a much larger, more turbinate-shaped shell with a higher, narrower spire,in having a much larger, more flaring aperture, and in having fewer spiral cords on the body whorl. *Homalopoma finkli* may belong to the subgenera *Leptothyropsis* Woodring, 1928 or *Cantrainea* Jeffreys, 1883.

Superfamily **Cypraeacea**
Family **Cypraeidae**
Genus *Cypraea* Linnaeus, 1758
Subgenus *Siphocypraea* Heilprin, 1887

Cypraea (Siphocypraea) mus tristensis new subspecies (Plate 22, Figures 5, 6, 9, 10)

Material Examined: HOLOTYPE - Length 48 mm, trawled by commercial shrimpers from 20 m depth of Tucacas, Golfo de Triste, Venezuela (11 42' N, 69 40' W), March 1979, USNM 784455; PARATYPES - length 46 mm, same location, depth, and dateas holotype, USNM; 5 adult specimens with preserved animals, collection of INTECMAR, Universidad Simon Bolivar marine laboratory, Puerto Cabello, Venezuela, trawled in Golfo de Triste of Tucacas, 1979.

Description: Shell ovate-elongate, dorso-ventrally flattened; aperture wide; base of shell flat; holotype with 17 dark brown teeth along inner edge of lip; columella with 12 dark brown teeth; beaks of anterior end developed into greatly-extended, flaring, ear-like projections; dorsum with incised central sulcus running from apex to anterior tip; color of dorsum pale bluish-white to light blue, with numerous well-spaced, pale tan spots; some specimens with pale brown bar-shaped markings along sides; dark brown patch on posterior end near apex; flattened base of shell tan to dark brown, with brown bars on outer lip, extending from tooth to edge of margin; pale dorsum color and darker base color meeting along flattened lateral margins; interior of aperture pale tan or yellow; living animal (frozen specimens examined at Simon Bolivar marine laboratory) uniformly bright tomato orange-red with numerous elongate, simple mantle papillae.

Etymology: Named for the Golfo de Triste, Venezuela, the type locality.

Discussion: *Cypraea mus tristensis* may eventually prove to be a distinct, full species in the *Cypraea mus-henekeni* complex. The characteristic pale blue-white dorsum color, flattened shell shape, and flaring ear-like anterior projections readily separate the more inflated, dark-colored *C. mus. Cypraea mus tristensis* also has a tomato-red animal with long papillae (illustrated in PETUCH, 1981, Figure 129), while *C. mus* has a tan-colored animal, with patches of dark brown, and numerous, white-tipped, short, spike-shaped mantle papillae (illustrated in color by BURGESS, 1986,p. 168).

Cypraea mus tristensis (illustrated in color in BURGESS, 1986, p. 270, figure H2 - as *"Cypraea mus")* appears to be a true, geographically-isolated subspecies of *C. mus.* As pointed out by BURGESS (1986, p. 168), *C. mus* has direct development, with no free-swimming veliger to disperse the gene pool. For this reason,the species appears to have a split range. Although *C. mus* was originally thought to extend from the Colombia-Panama border (Golfo de Uraba) to east of the Paraguana Peninsula, Venezuela (BURGESS, 1986, p. 168; PETUCH, 1979), more recent studies done by myself have shown this to be in error. *Cypraea mus mus* is confined to intertidal and shallow water Turtle Grass *(Thalassia)* beds, from Riohacha, Colombia, around the Goajira Peninsula, throughout the Gulf of Venezuela,and along the western and northern coasts of the Paraguana Peninsula, Venezuela. The eastern coast of the Paraguana Peninsula and the coastlines of eastern Falcon State and Carabobo State are composed of open, sandy beaches that are exposed to constant, extremely heavy surf. This heavy surf area, which is devoid of *Thalassia,* acts as a geographical barrier between the quiet, protected waters of the widely-separated Gulf of Venezuela and Gulf of Triste. *Cypraea mus tristensis* probably became separated and isolated from the main *C. mus* gene pool when sea levels rose to their present height about 25,000 years ago. This would have flooded the shallow coastline along Falcon and Carabobo States, producing the sand shoals that now act as a biotic barrier to the shallow water faunas of the Gulfs of Venezuela and Triste.

For the sake of clarity,it should be pointed out here that, to the best of my knowledge, *Cypraea donmoorei* Petuch does indeed, represent a full species distinct from *C. mus.* Besides being a much larger shell (64-70 mm on average, as opposed to 30-35 mm for *C. mus), C. donmoorei* has a paler color pattern composed of tan spots on a bluish-white background, well-developed, very coarse teeth on the columella, a unique, large bright orange columellar patch, a narrower aperture, and a white interior instead of a dark brown as in *C. mus.* The juvenile, bulla stages of the two species are also very different; that of *C. donmoorei* (Plate 22, Figure 13) is pale blue-white with large, prominent dark purple zig-zags, while that of *C. mus* (Plate 22, Figure 14) is dark brown with close-set dark brown maculations. The animals and ecologies of the two species are completely different. The animal of *C. mus* is described above, and it is an herbivore, feeding on diatoms that encrust the *Thalassia* blades, and on pieces of the *Thalassia* itself. *Cypraea donmoorei* is a deep water species that feeds on sponges and possibly hydrozoans, and has a completely white animal with long, shaggy papillae that make the living animal resemble a Hostess snowball cupcake. A preserved animal of *C. donmoorei* is deposited in the Invertebrate Museum, Rosenstiel School of Marine and Atmospheric Science, University of Miami, and is readily accessible for examination. The animals of the three species, *C. mus, C. mus tristensis* (as *C. henekeni),* and *C. donmoorei,* were illustrated and discussed in detail by myself (PETUCH, 1981), but have been ignored by subsequent workers.

Cypraea donmoorei, like *C. mus tristensis,* is a rare species, seldom seen in collections. Shell dealers in the United States and Europe have recently been selling large, light-colored *C. mus* as *C. donmoorei,* adding to the confusion regarding the two species. Virtually every specimen of *"C. donmoorei"* that I have examined in private collections and shell shows has turned out to be *C. mus,* and not my species. Also adding to the confusion has been a recent misdesignation of the holotype of *C. donmoorei.* BURGESS, 1986 (p. 270, Figure H1) illustrates in color a specimen that he states is the holotype of *C. donmoorei.* This photograph was apparently sent to him by Mr. Rosewater of the USNM, who photographed the wrong specimen. The real holotype is illustrated here (Plate 22, Figure 7, 8). Collectors should beware of buying or trading specimens of *C. mus* labeled as *C. donmoorei.*

In being cautious, I originally referred to *C. mus tristensis* as *C. henekeni* Sowerby, a Pliocene Caribbean fossil. Mr. BURGESS (1986, p. 270, figure H3) illustrates a specimen of *C. henekeni,* and as can be seen, it differs from *C. mus tristensis*. For this reason, I have proposed a new subspecific taxon.

Family **Triviidae**
Genus *Discotrivia* Cate, 1979

Discotrivia labiosa (Gaskoin, 1836) (Plate 22, Figures 15, 16)

Material Examined: Length 19 mm, collection of Rosenstiel School of Marine and Atmospheric Science, University of Miami,trawled by *R/V John Elliot Pillsbury* from 100 m depth off Cabo La Vela, Goajira Peninsula, Colombia, 1966.

Discussion: At 19 mm in length, *Discotrivia labiosa* is the largest trivia known from the western Atlantic. Although the type locality was unknown and subsequent workers, such as CATE (1979), thought it might be a Panamic species, the recently-collected Colombian specimen demonstrates that the species is actually part of the Caribbean fauna. This beautiful, large species needed to be reillustrated.

Family **Ovulidae**
Genus *Pseudocyphoma* Cate, 1973

Pseudocyphoma gibsonsmithorum new species (Plate 26, Figures 6, 7)

Material Examined: HOLOTYPE - Length 15 mm, width 6 mm, found on beach, Borboruta, Carabobo State, Venezuela, 1976, USNM 859829; PARATYPE - 1 specimen, same locality as holotype, Gibson-Smith collection.

Description: Shell small for genus, thin, fragile, translucent; general shell form elongated, slender, with projecting terminals; outer lip very thickened; dorsum with numerous very fine, incised spiral grooves, giving shell silky appearance; base of shell smooth and shiny; aperture narrow, curved, becoming wider at anterior end; columellar side of posterior extremity with single large, protruding plication; posterior one-half of shell more slender than anterior half; shell color uniformly cream-white; interior of aperture cream-white; dorsum with weakly-developed transverse ridge; transverse ridge white in color, contrasting faintly with rest of shell.

Etymology: Named for Dr. Jack and Mrs. Winifred Gibson-Smith, Caracas, Venezuela, who collected the holotype.

Discussion: *Pseudocyphoma gibsonsmithorum* is closest to *P. intermedium* (Sowerby), a widespread Caribbean species. The new Venezuelan species differs from *P. intermedium* by being a smaller, more elongated shell, by having shorter extremities, by having a smaller, more oblique, plication on the columellar side of the posterior extremity, and in having the fine incised grooves on the dorsum. To date, *P. gibsonsmithorum* has only been collected along the Carabobo coast of Venezuela, in the vicinity of the Golfo de Triste.

Genus *Phenacovolva* Iredale, 1939
Subgenus *Turbovula* Cate, 1973

Phenacovolva (Turbovula) lindae new species (Plate 26, Figures 8, 9)

Material Examined: HOLOTYPE - Length 17 mm, width 7 mm, trawled by commercial shrimpers from 50 m depth off Cabo La Vela, Goajira Peninsula, Colombia, 1976, USNM 859828.

Description: Shell thin, fragile, inflated, with projecting terminals; outer lip thickened, well-developed; posterior extremity with large plication along margin of aperture; body whorl with silky texture; aperture narrow, widening and flaring at anterior end; fossula well-developed, projecting from columella; inner edge of lip with very weak crenulations; color uniformly pale cream-white.

Etymology: Named for my wife Linda J. Petuch.

Discussion: *Phenacovolva lindae* is the second-known member of this primarily Japonic-Australian subgenus found in the Caribbean Sea. The other Caribbean species, *P. (Turbovula) bartschi* Cate, is only known from deep water off the northern coast of Puerto Rico. *Phenacovolva lindae* differs from *P. bartschi* by being a larger, more inflated shell, by having large, more projecting extremities (especially the posterior extremity), by having a large plication on the columellar portion of the posterior extremity, and by having a much narrower aperture. Antipatharian octocorallians were commonly dredged from the offshore areas of Cabo La Vela along the Goajira Peninsula, and these may serve as the host species for *P. (Turbovula) lindae.*

Family **Cassidae**

Genus *Morum* Röding, 1798

Subgenus *Cancellomorum* Emerson and Old, 1963

Morum (Cancellomorum) lindae new species (Plate 23, Figures 1, 2)

Material Examined: HOLOTYPE - Length 35 mm, trawled by *R/V John Elliot Pillsbury* (University of Miami research vessel) from 11 m depth off Cabo La Vela, Goajira Peninsula, Colombia (station p-772), 1966, USNM 859831.

Description: Shell stocky, rotund, inflated, lightweight; spire low, but early whorls exserted; shoulder angle sharp; 16 serrated, bladelike axial ribs on body whorl; axial ribs intersected by 7 large spiral cords, with large, pointed serration at each intersection, giving shell squamose appearance; serrations along shoulder cord largest, producing coronated effect; outer lip thickened; inner edge of lip with 16 large denticles and smaller secondary denticles in between; parietal shield well-developed, wide, covered with numerous small pustules; shell base color pinkish-white with scattered patches of bright pink; pinkish-white base color overlaid with 2 wide, reddish-brown spiral bands; spire with intermittent large reddish-brown patches; parietal shield salmon pink with lavender border and white pustules; side of lip white, with numerous bands of tiny brown dots; edge of lip salmon-pink with 13 lavender-purple radiating bars; denticles of lip salmon-pink; interior of aperture white; protoconch and early whorls white.

Etymology: Named for my wife, Linda J. Petuch.

Discussion: Along with *Morum dennisoni* (Reeve) and *M. matthewsi* Emerson, *M. lindae* is the third known member of the subgenus *Cancellomorum* to be found in the western Atlantic. I originally referred this species to the Pliocene Caribbean *M. dominguense* (Sowerby) (PETUCH, 1981), but I now feel that the living species is different enough to deserve a name of its own. The fossil *M. dominguense,* although similar to *M. lindae,* has fewer axial ribs, is broader and more angled, has a smaller parietal shield, and is far less sculptured and less squamose. *Morum dominguense,* however, is almost certainly the direct ancestor of *M. lindae.* The only other closely-related living species, *M. dennisoni* (Plate 23, Figures 3, 4), differs from *M. lindae* in being a much larger, more elongated shell with a sloping shoulder, in having a much wider parietal shield that is colored, as is the lip, bright orange-red with white pustules, and in having finer and more numerous axial ribs and spiral cords. In general, *M. dennisoni* is a much less ornate shell than *M. lindae,* lacking the large serrated fimbriations of the squamose *M. lindae.* Bayer (1971, Figure 16) illustrated this new species but considered it a variety of *M. dennisoni.* To date, *M. lindae* is known only from the unique holotype, making it, because of its great beauty, one of the most desirable shells from the Caribbean.

Genus *Sconsia* Gray, 1847

Sconsia lindae new species (Plate 15, Figures 7, 8)

Material Examined: HOLOTYPE - Length 71 mm, trawled by commercial shrimp boat from 35 m depth off Cartagena, Colombia, 1976, USNM 784457; PARATYPES - lengths 49 mm and 51 mm, trawled by shrimpers from 37 mm depth of Gulf of Uraba, Colombia, invertebrate museum collection RSMAS, University of Miami; length 43 mm, 25 m depth off Gulf of Uraba, Colombia, RSMAS collection; one specimen, Robert Pace collection.

Description: Shell inflated, rounded, thickened; form slightly dorso-ventrally flattened; spire somewhat elevated; body and spire whorls smooth, shiny, sculptured with numerous fine, raised, spiral threads, giving shell silky appearance; dorsum of lip with prominent axial ribs which intersect with spiral threads to produce beaded appearance; shell color white with variable amounts of red-brown mottlings, usually in form of vertical flammules in zebra-like pattern; color pattern may vary from pure white to having 4-6 rows of brown checkers on alternating vertical flammules composed of coalesced checkered bands; spire white, often with band of brown flammules; interior of aperture white; outer lip of shell developed, thickened, porcelaneous white.

Etymology: Named for my wife, Linda J. Petuch.

Discussion: I originally referred this new species of *Sconsia* to the Pliocene fossil species *S. laevigata* Sowerby (Petuch, 1981, pp 321-322). Dr. Emily VOKES of Tulane University (in litt.) has stated that a fossil specimen of *S. laevigata* examined by her under ultraviolet light revealed a color pattern composed of four bands of large checkers, and not the zebra pattern seen in *S. lindae.* In many ways, the fossil *S. laevigata* more closely resembles the living *S. nephele* Bayer, (Plate 23, Figure 5), from deep water off the Lesser Antilles, and is probably the ancestor of that species. Because of the zebra color pattern and elongated shape, I have decided to separate the living Colombian species from the Pliocene fossil. *Sconsia laevigata,* however, certainly is the ancestor of all western Atlantic *Sconsia* species.

Sconsia lindae differs from the widespread *S. striata* (Lamarck) in being a more elongated shell, in having finer sculpturing and a silky texture, and having a zebra color pattern instead of rows of checkers. Ecologically, the two species are also quite different; *S. striata* is a deep water species (av. depth 200 m) found on the continental slope or edge of the continental shelf, while *S. lindae* is found in fairly shallow depths (av. 30 m) on the continental shelf. At Santa Marta, Colombia, *S. lindae* has been collected by skin divers (snorkeling) in only 5 m depth.

Family **Muricidae**
Genus *Dermomurex* Monterosato, 1890
Subgenus *Trialatella* Berry, 1964

Dermomurex (Trialatella) kaicherae new species (Plate 24, Figures 17, 18)

Material Examined: HOLOTYPE - Length 16 mm, width 9 mm, trawled by commercial shrimpers from 35 m depth off Cabo La Vela, Goajira Peninsula, Colombia, 1977, USNM 859837.

Description: Shell elongated, fusiform, thin and fragile; 3 thin, winglike varices per whorl; body whorl and spire whorls ornamented with 3 large, evenly-spaced spiral cords; intervarical regions with an axial rib composed of 3 large knobs; 3 large cords extend onto backs of winglike varices; entire shell covered with numerous fine spiral threads, giving shell a rough appearance; aperture large, oval, with 6 tiny red denticles along inner edge of lip; entire shell uniform chocolate-brown; interior of aperture brown; columella tan; intritacalx (eroded off much of holotype) tan.

Etymology: Named for Mrs. Sally D. Kaicher of St. Petersburg, Florida, a noted authority on the mollusca of the Caribbean.

Discussion: *Dermomurex kaicherae* is the fourth-known member of the subgenus *Trialatella* to be discovered in the western Atlantic. Of these species, *D. kaicherae* is closest to *D. abyssicola* (Crosse) from the Lesser Antilles, but differs in being a more slender, elongated shell with a higher spire, in being a rougher-textured shell with prominent intervarical, axially-arranged knobs, and in lacking the dark brown band at the base of the body whorl. Unlike *D. abyssicola,* the entire shell of *D. kaicherae* is brown. *Dermomurex kaicherae* is also similar to *D. oxum* Petuch from the Abrolhos Islands off south central Brazil, but differs in being a more-slender, higher-spired species, and in having an all-brown shell color as opposed to white in *D. oxum* (Plate 27, Figures 10, 11). *Dermomurex kaicherae* is a rare species, known only from the holotype taken off the Goajira Peninsula of Colombia.

Subfamily **Muricinae**

Genus *Lindapterys* new genus

Diagnosis: Shells small for family, dorso-ventrally flattened; 2 winglike varices per whorl, with varices aligned up the spire; varices pointed, distinctly channeled; last varix with channel connected to aperture as anal channel; numerous thin, sharp axial ribs between varices, running from suture to siphonal canal; siphonal canal short, open; columellar region with large, adherent parietal shield; outer edge of lip flaring, extending beyond varix; parietal shield and flaring lip producing peristome; inner edge of lip with several denticles; shell sculptured with numerous raised spiral threads; protoconch tiny, similar to genus *Pterochelus;* first postnuclear whorls with several cord-like ribs.

Type Species: *Lindapterys vokesae* new species, described herein, Chipola Formation, Miocene of northern Florida. Illustrated and discussed by E. VOKES (1974, pp. 96-98).

Species in *Lindapterys: Lindapterys poppelacki* (Hörnes, 1853), Miocene of Austria; *Lindapterys sanderi* Petuch, described herein, Recent of Barbados; *Lindapterys vokesae,* described herein, Chipola Foundation, Miocene of northern Florida.

Etymology: The taxon is a combination of the words "Linda" (for my wife, Linda J. Petuch) and "Pterys" (wings), in reference to the two winglike varices per whorl. The taxon is neuter.

Discussion: Until this description, this unusual group of muricids was known only from a few fossil specimens from two locations. The first-known *Lindapterys* species was collected in the rich Miocene fossil beds of the Leithakalk, about 50 miles north of Vienna, Austria by Hörnes in the early 1800's. Because of the aligned two varices per whorl, Hörnes placed his species in the bursid genus *Ranella.* In 1974, Dr. Emily Vokes of Tulane University described and illustrated the first-known American specimen of *Lindapterys* (VOKES, 1974, pp. 96-98), which had been collected in the Miocene Chipola Formation of northern Florida. Being similar-appearing to the Vienna Miocene specimens, Dr. Vokes referred the single-known Chipola specimen to Hörnes' *Ranella poppelacki.* In 1979, Dr. Finn Sander of the McGill University Marine Laboratory, St. James, Barbados, showed me a Recent specimen of *Lindapterys* that was dredged during his research cruises from deep water along the Barbados seamount. This specimen confirms that *Lindapterys* is a living component of the Caribbean deep water fauna. At present, the genus is represented by a few specimens from Austria, a single specimen from Florida, and a single specimen from Barbados. The fact that no specimens are known in the fossil record, from the early Miocene to the Holocene, is problematical indeed. The single living species from Barbados stands alone as an enigma.

Lindapterys, although at first appearing more like a cymatiid or bursid, does fit logically into the plethora of genera that comprises the Muricidae. It combines the characteristics of two genera: *Pterochelus* and *Aspella* s.s. With *Pterochelus, Lindapterys* shares the open anal canal that crosses over the posterior end of the varix. Although having three varices per whorl, small species of *Pterochelus,* such as *P. angasi* (Crosse), *P. eos* (Hutton), and *P. ariomus* (Clench and Farfante), closely resemble *Lindapterys.* In only having two varices per whorl and in having distinct denticles on the inner edge of the lip, *Lindapterys* also resembles *Aspella* s.s. species such as *A. hastula* (Reeve) and *A. pyramidalis* (Broderip). Unlike *Aspella, Lindapterys* lacks a chalky intritacalx, and in this manner, appears to be closer to *Pterochelus.*

Lindapterys sanderi new species (Plate 15, Figures 5, 6)

Material Examined: HOLOTYPE - Length 11 mm, width 6 mm, trawled by McGill University research vessel from 300 m depth off St. James, Barbados, 1979, USNM 859838.

Description: General morphology as for genus; 6 axial ribs between varices; 5 labial denticles; outer edge of lip very flaring, almost obscuring the underlying varix.

Etymology: Named for Dr. Finn Sander, McGill University Marine Laboratory, St. James, Barbados, who collected the holotype.

Discussion: *Lindapterys sanderi* is closest to *L. vokesae* Petuch (following description) from the Chipola Formation of Florida, but differs in being stumpier and less elongated, in having a shorter siphonal canal, and by having a larger, far more flaring lip. *Lindapterys sanderi* is most probably the direct descendant of *L. vokesae.*

Lindapterys vokesae new species (Plate 15, Figures 10, 11)

Material Examined: HOLOTYPE - Length 11.7 mm, width 5.8 mm, locality TU951 (Tulane University Locality), Ten Mile Creek, Calhoun County, Florida, Chipola Formation, early Miocene of North Florida, USNM 647012 (Type collection of Dept. of Paleobiology).

Description: General morphology as for genus; 6 axial ribs between varices; 5 tiny labial denticles; outer edge of lip flaring but narrow.

Etymology: Named for Dr. Emily Vokes of Tulane University, New Orleans, in recognition of her pioneer work on the Muricidae of the western Atlantic, and who collected the holotype.

Discussion: Although a fossil species, I felt it was important to list and describe this interesting little shell, the ancestor of the Barbados *Lindapterys sanderi,* within this publication on living Caribbean mollusks. A comparison of *L. vokesae* with the living *L. sanderi* was given under the last description. *Lindapterys vokesae* is also similar to the contemporaneous Austrian Miocene *L. poppelacki* (Hörnes), but differs in having fewer axial ribs between the varices (6 as opposed to 8), in having larger, more wing like varices, in having smaller denticles on the edge of the lip, and in having a narrower edge to the flaring outer lip.

Genus *Murex* Linnaeus, 1758

Murex sunderlandi new species (Plate 21, Figures 5, 6)

Material Examined: HOLOTYPE - Length 41 mm, width 21 mm, trawled by commercial shrimpers from 35 m depth off Cabo La Vela, Goajira Peninsula, Colombia, 1979, USNM 859841; PARATYPES - 1 specimen, same locality and depth as holotype, Kevan Sunderland collection; 1 specimen, same locality as holotype, Robert Pace collection.

Description: Shell very thin, lightweight, with high, protracted spire and long, straight siphonal canal; shell with 3 varices per whorl; varices thin, low, with 3 thin spines; spine on shoulder longest; siphonal canal with 1 short spine near junction with body whorl; aperture large, flaring; parietal shield large, erect, nonadherent; shell color pale tan overlaid with 8-10 reddish-brown spiral lines; intervarical regions with 5 low, faintly nodose axial ribs; areas between spines on varices with numerous reddish-brown ribs; ribs correspond to spiral lines of intervarical regions; interior of aperture white, with reddish-brown lines showing through.

Etymology: Named for Mr. Kevan Sunderland of Ft. Lauderdale, Florida.

Discussion: *Murex sunderlandi* is most similar to the widespread, deep water Caribbean *M. tryoni* Hidalgo, but differs in being a more inflated, thinner shell with a higher spire, and in having smoother, more rounded intervarical regions with the axial ribs being greatly reduced or absent. The greatest difference between these two species is seen in the color pattern; *M. tryoni* is a pure white shell with three widely-spaced brown lines, while *M. sunderlandi* is a tan shell with 8-10 dark reddish-brown lines. With a brown-lined color pattern and three spines per varix, *M. sunderlandi* closely resembles the West Mexican *M. tricoronis* Berry. *Murex sunderlandi* may be the Atlantic analogue species of the Panamic *M. tricoronis.* At present, *M. sunderlandi* is known only from the Goajira Peninsula area of northern Colombia.

Genus *Murexiella* Clench and Farfante, 1943

Murexiella hilli new species (Plate 17, Figures 6, 7)

Material Examined: HOLOTYPE - Length 34 mm, on yellow mussel bed *(Bachidontes modiolus),* low tide, Malmok, Aruba Is., Netherlands Antilles, 1983, USNM 859842; PARATYPES - 3 specimens, same locality as holotype, Leonard Hill collection; 2 specimens same locality as holotype, Robert Pace collection; 2 specimens, same locality as holotype, Kevan Sunderland collection.

Description: Shell large for genus, globose, with inflated whorls and rounded shoulder; spire moderately elevated; siphonal canal long, straight; aperture large in proportion to shell size, oval; 8 recurved varices per whorl; body whorl ornamented with 5 large, fimbriated cords between varices; large, recurved, fimbriated spine on varix where intersected by cord; subsutural area flattened, producing stepped spire; spire whorls with 2 fimbriated cords; siphonal canal with 3 large, fimbriated and branching curved spines; 2 smaller spines between 3 siphonal spines; siphonal spines curved posteriorward; color varying from white to pink, salmon (holotype), lavender and dark brown; brown band around subsutural area; siphonal canal generally white.

Etymology: Named for Mr. Leonard Hill, of Miami, Florida, who kindly donated the holotype.

Discussion: *Murexiella hilli* has been passed amongst shell collectors and museums, for years now, under the name of *Murexiella facetus* Vokes. That species, however, is a much smaller, quite different-looking fossil shell, known only from the Pliocene shell beds of southern Florida. Why the name of a small Pliocene fossil species from Florida was applied to this large, ornate species from Aruba is indeed a mystery. *Murexiella hilli* has also been called a variety of *M. macgintyi* (M. Smith), a small species from Florida, northern Cuba, and the Bahamas. The new Aruban species, however, differs from *M. macgintyi* by being a much larger, more globose shell with rounded varices (instead of angled as in *M. macgintyi),* by having more varices per whorl (8 as opposed to six), by having larger, more fimbriated and recurved varical and siphonal spines, and by having a longer siphonal canal with more siphonal spines. *Murexiella hilli* is also a more variable shell when it comes to shell color, often exhibiting lilacs, pinks, salmon-pinks, and brown, while *M. macgintyi* tends to be conservative in color being mostly tans or pale orange. *Murexiella hilli* is endemic to Aruba, being found primarily along the Malmok coast on mussel beds. For a discussion of distribution of the three species in the

Murexiella macgintyi complex, see the description of *Murexiella leonardhilli* under the Brazilian section (New Discoveries From Brazil).

Interestingly enough, the type of *Murexiella macgintyi* is also a fossil from the Plio-Pleistocene shell beds of southern Florida. The living species from Florida and the Bahamas, although traditionally called *M. macgintyi,* looks somewhat different when compared to the fossil holotype. The living species is larger with more rounded varices. Further study may show that it may actually represent an unnamed species that evolved from the ancestral Pliocene *M. macgintyi.*

Genus *Murexsul* Iredale, 1915

Murexsul harasewychi new species (Plate 21, Figures 11, 12)

Material Examined: HOLOTYPE - Length 15 mm, width 8 mm, trawled by commercial shrimpers from 35 m depth off Cabo La Vela, Goajira Peninsula, Colombia, 1977, USNM 859845; PARATYPE - 1 specimen, same locality and depth as holotype, Robert Pace collection.

Description: Shell small for genus, thin, fragile, very elongated and fusiform, with high elevated spire; siphonal canal long, roughly one-third total length; 6 varices per whorl; shoulder sharply angled; body whorl sculptured with 5 large, fimbriated cords; area between suture and shoulder without large cords but with numerous fine spiral threads; shoulder of each varix with 1 small, bifurcated spine; 5 spiral cords overlap onto varices, producing low, fimbriated knobs; varices on siphonal canal with 2 short spines; aperture round-oval, with raised peristome; inner edge of lip without denticles; shell color two-toned, with body whorl and spire being yellow-orange and with siphonal canal being brown; early whorls, protoconch, and tips of shoulder spines pale reddish-brown; interior of aperture and peristome white.

Etymology: Named for Dr. M. G. Harasewych, Division of Mollusks, National Museum of Natural History, Smithsonian Institution.

Discussion: *Murexsul harasewychi* is the first member of the genus to be collected in the western Atlantic. This new species is closest to the eastern Atlantic and Mediterranean *M. aradasii* (Monterosato), but differs in being a more slender, elongated shell with fewer varices, in having a smaller aperture and proportionally longer siphonal canal, and in having the two-toned color pattern. This first-known Caribbean *Murexsul* has only been collected off the Goajira Peninsula of Colombia.

Genus *Muricopsis* Bucquoy, Dautzenberg, and Dollfuss, 1882

Muricopsis aliceae new species (Plate 17, Figures 8, 9)

Material Examined: HOLOTYPE - Length 11 mm, found dead, in coral rubble, 2 m depth, on reef off southern coast of Bonaire Is., Netherlands Antilles, 1980, USNM 859846; PARATYPES - 2 specimens, same locality and depth as holotype, Robert Pace collection.

Description: Shell small for genus, elongated, with high, elevated spire; 6 fimbriated varices per whorl; varices with 6 short, bladelike serrations; intervarical areas and backs of varices of body whorl with 6 large, raised, fimbriated spiral cords; inside of lip with 6 large denticles; shell color pure white (holotype encrusted with the red foraminiferan *Homotrema rubrum).*

Etymology: Named for Mrs. Alice Pace, who helped collect the type lot on Bonaire.

Discussion: *Muricopsis aliceae* is closest to *M. richardbinghami* Petuch from Florida, but differs in being a smaller shell with a much shorter siphonal canal, in having smaller varical serrations and less ornate sculpturing, and in being pure white as opposed to the tan and brown-banded color of *M. richardbinghami. Muricopsis aliceae* lives in the same habitat as *Conus aurantius* Hwass, and like that species may be endemic to the islands of Curaçao and Bonaire. At present, *M. aliceae* is known only from the type locality. *Muricopsis tulensis* Radwin and D'Attilio from the Gulf of California is similar to *M. aliceae,* both in size and shape, and may be the Pacific analogue species.

Family **Fasciolariidae**
Subfamily **Fusininae**
Genus *Harasewychia* new genus

Diagnosis: Shell small for subfamily, fusiform in shape, thin delicate, fragile; spires elevated, usually scalariform, with tabulate whorls; shells characteristically ornamented with sharp axial ribs overlaid with large, evenly-spaced spiral cords; spiral cords and axial ribs intersect to produce row of small knobs down length of axial rib; protoconch large, mamillate, smooth, usually composed of 2 or 2 1/2 rounded whorls; shell color pure white; some species being vitreous and translucent; periostracum usually thick, velvety, and tightly adherent; ecologically, a deep water group.

Type Species: *Harasewychia harasewychi* new species, described herein, 200 m depth off Monges Islands, Golfo de Venezuela, Venezuela.

Species in *Harasewychia: Harasewychia amiantus* (Dall), deep water, Gulf of Mexico; *Harasewychia benthalis* (Dall), deep water, Florida Strait; *Harasewychia alcimus* (Dall), deep water, Gulf of Mexico; *Harasewychia aepynotus* (Dall), deep water, Gulf of Mexico, Florida Strait; *Harasewychia rushii* (Dall), deep water eastern Florida; *Harasewychia amphiurgus* (Dall), deep water, Gulf of Mexico; *Harasewychia vitreus* (Dall), deep water, Georgia; *Harasewychia harasewychi.*

Etymology: Named for Dr. M. G. Harasewych of the Smithsonian Institution.

Discussion: This morphologically close-knit group of small, fragile, deep water fasciolariids has generally been referred to the genus *Barbarofusus* Grabau and Shimer, 1909 by most recent workers. *Barbarofusus* (as characterized by *B. harfordi* (Stearns) and *B. kobelti* (Dall)), however, is a Californian and Panamic genus which typically contains larger species with minutely fimbriated cords and dark brown or orange colors. *Harasewychia* differs from *Barbarofusus* by having thinner, more delicate shells (often translucent), by having smaller shells, in having axial rows of beads instead of fimbriated cords, and in being universally white instead of colored. As far as is known, no species of *Harasewychia* are found in the eastern Pacific. The genus is apparently restricted to deep water (outer continental shelf-slope) areas from Georgia south to Venezuela, with the majority of species being concentrated within the Gulf of Mexico. In the form of shell sculpture, *Harasewychia* converges on the Japonic-Western Pacific genus *Granulifusus.*

Harasewychia harasewychi new species (Plate 13, Figures 9, 10)

Material Examined: HOLOTYPE - Length 34 mm, width 15 mm, trawled by commercial deep water shrimpers from 200 m depth, near Los Monges Islands, mouth of the Gulf of Venezuela, Venezuela, 1975, USNM 859852; PARATYPES - 1 specimen, same locality and depth as holotype, Kevan Sunderland collection; 1 specimen, same locality and depth as holotype, Robert Pace collection.

Description: General shell form and color as for genus; shoulder and spire whorls sharply-angled, giving shell tabulate appearance; periphery of shoulder with one, strong, carina-like cord; 15-17 thin, evenly-spaced axial ribs per whorl; ribs overlaid by 10 large spiral cords; large, knoblike bead produced at intersection of axial and spiral cords; fine spiral threads between large spiral cords and on siphonal canal; sloping area between shoulder cord and suture without coarse sculpture.

Etymology: Named for Dr. M. G. Harasewych of the Smithsonian Institution.

Discussion: *Harasewychia harasewychi* is closest to *H. benthalis* (Dall), both in size and sculpture pattern, but differs in having sharply-angled whorls and a tabulate spire, in having the relatively unsculptured, smooth area between the shoulder cord and the suture, and in having more numerous axial ribs and spiral cords. *Harasewychia harasewychi* occurs in the same cold, upwelling areas as does *Mohnia kaicherae* Petuch. This may explain why a normally deep water group (average 300 m depth) such as *Harasewychia* was collected in only 200 m depth off the mouth of the Gulf of Venezuela.

Family **Colubrariidae**
Genus *Colubraria* Schumacher, 1817

Colubraria bayeri new species (Plate 24, Figures 11, 12)

Material Examined: HOLOTYPE - Length 17 mm, width 8 mm, trawled by commercial shrimpers from 35 m depth of Cabo La Vela, Goajira Peninsula, Colombia, 1975, USNM 859853.

Description: Shell small for genus, very thin and fragile, inflated, bullate; spire low in comparison to cogeners; shoulder and spire tabulate; 2 very thin, bladelike, low varices per whorl; 16-18 very narrow, low, axial riblets between sets of varices; axial riblets overlaid with numerous raised spiral threads; parietal shield erect, nonadherent; edge of lip thin, bladelike, crenulated; color pale tan with numerous large, scattered, amorphous reddish-brown patches; varices marked with one large medial reddish-brown checker-shaped flammule and smaller brown dashes; interior of aperture pale tan, with brown checkers of labial varix showing through.

Etymology: Named for Dr. Frederick M. Bayer, Dept. of Invertebrates, National Museum of Natural History, Smithsonian Institution, in recognition of his pioneer research on the mollusks of the southern Caribbean.

Discussion: Of the western Atlantic Colubrariidae, *Colubraria bayeri* appears closest to *C. monroei* McGinty, from the northern coast of Cuba. Although both species are small, fragile shells, *C. bayeri* differs from *C. monroei* in being a much more inflated shell with a lower spire, and in having a much wider, almost humped, body whorl. *Colubraria monroei* is a pale whitish-tan shell, while *C. bayeri* is darker-colored, being tan with large reddish-brown blotches. At present, *C. bayeri* has only been collected off the Goajira Peninsula of Colombia.

Family **Buccinidae**
Genus *Antillophos* Woodring, 1928

Antillophos bayeri new species (Plate 24, Figures 4, 5)

Material Examined: HOLOTYPE - Length 17 mm, width 7 mm, trawled by commercial shrimp trawler from 35 m depth off Cabo La Vela, Goajira Peninsula, Colombia, 1975, USNM 859854; PARATYPE - 1 specimen, same locality and depth as holotype, Robert Pace collection.

Description: Shell small for genus, very elongated with protracted, scalariform spire; whorls with 3 or 4 large, thin varices; final varix wide, thickened; areas between varices with finely reticulated pattern of intersecting axial and spiral threads; siphonal canal short, distinctly constricted; aperture small, narrow; columella with 5 large denticles; aperture lirate within; anterior end of lip with shallow stromboid notch; color pale tan with 3 reddish-tan spiral bands; varices white with 3 darker reddish-tan checkers that correspond to spiral bands; columella and interior of aperture white.

Etymology: Named for Dr. Frederick M. Bayer, Dept. of Invertebrates, NMNH, Smithsonian

Institution.

Discussion: *Antillophos bayeri* is most similar to the widespread *A. candei* (Orbigny) (Plate 24, Figure 6), but differs in being a much smaller, much more slender shell with finer sculpturing, and in having more numerous, large, thin varices on the body whorl and spire. In this last character, *A. bayeri* somewhat resembles a small colubrariid. The siphonal canal of *A. bayeri* is also more constricted than that of *A. candei.* Of the species of *Antillophos, A. bayeri* is the only one to have a well-developed stromboid-type notch on the anterior end of the lip. *Antillophos bayeri* is a rare species, apparently being restricted to the upwelling system off the Goajira Peninsula of Colombia.

Genus *Mohnia* Friele, 1878

Mohnia kaicherae new species (Plate 21, Figures 8, 9)

Material Examined: HOLOTYPE - Length 29 mm, width 12 mm, trawled by deep water shrimp trawler from 200 m depth off Los Monges Islands, off mouth of Gulf of Venezuela, Venezuela, 1975, USNM 859855.

Description: Shell large for genus, fusiform, with elevated spire; whorls sloping, with slightly-angled shoulders; holotype with 12 widely-spaced, rounded axial ribs around shoulder; entire shell covered with large, evenly-spaced spiral threads, giving shell rough texture; siphonal canal well-developed; inside of lip with 14 large, elongated denticles; denticles disappear farther in aperture; columella narrow, smooth; color uniform cream-white.

Etymology: Named for Mrs. Sally D. Kaicher of St. Petersburg, Florida, in recognition of her many contributions to systematic malacology.

Discussion: *Mohnia kaicherae* is the first species of the primarily arctic (Atlantic and Pacific) genus to be found in the Caribbean Sea. Previously, the southernmost *Mohnia* species was *M. carolinensis* (Verrill), which ranges from North Carolina to the Florida Keys in deep water. *Mohnia kaicherae* is similar to *M. carolinensis,* but differs in being a larger shell with a more angled shoulder, in being a more elongated shell with a higher spire, and by having finer and more numerous spiral threads around the shell.

Mohnia kaicherae is also found in shallower water than most of the other *Mohnia* species. A large upwelling system exists just off the mouth of the Gulf of Venezuela, producing a cold water environment and allowing deepwater or Arctic-type mollusks, such as *Mohnia,* to move up into shallow shelf areas.

Genus *Truncaria* Adams and Reeve, 1850

Truncaria lindae new species (Plate 21, Figures 16, 17, 18)

Material Examined: HOLOTYPE - Length 16 mm (fragment), width 10 mm, trawled from 65 m depth, off Cabo Cordera, Venezuela, by *R/V John Elliot Pillsbury,* University of Miami, 1966, USNM 859856; PARATYPE - Length 13 mm (fragment) same locality and depth as holotype, USNM 859857.

Description: (Based on fragments) Shell thin, fragile, bulliform, inflated; reconstructed outline (based on related species and general outline of fragments) of shell elongated, with high spire; aperture wide, flaring at anterior end, producing open anterior canal; shell sculptured with numerous fine spiral riblets and threads; outer lip smooth, without denticles along inner edge; columella with adherent parietal shield; anterior end of columella abruptly terminated; shell color yellow-tan with large cloudings of darker tan; yellow-tan base color overlaid with large, evenly-spaced, dark brown lines; interior of aperture yellow-tan.

Etymology: Named for my wife, Linda J. Petuch.

Discussion: The finding of a Caribbean *Truncaria* species is an important discovery, especially since the genus was previously known from only one rare species from the Bay of Panama (Pacific Ocean), *Truncaria filosa* (Adams and Reeve) (=*Truncaria brunneocincta* Dall). The new Caribbean species differs from the Pacific *T. filosa* in being (as extrapolated from the fragments) a slightly stockier shell, and by having a different color pattern. *Truncaria lindae* is yellow with tan cloudings and dark brown spiral lines, while *T. filosa* is bright pink with numerous fine brown spiral lines.

Although this new taxon is only represented by two fragments, enough of the body whorl and apertural region remain to allow for a formal description. The flaring aperture, distinctly truncated columella, and sculpture pattern of fine brown lines, all show that the fragments belong to the genus *Truncaria*. I felt it was important that this new species be described, even from fragments, in order to establish a Caribbean member of the genus, and to remove *Truncaria* from the list of paciphilic genera. OLSSON (1971, p. 62) apparently saw these two fragments in the University of Miami collection and commented on the possibility of there being a Caribbean *Truncaria*.

Family **Olividae**
Subfamily **Ancillinae**
Genus *Gracilancilla* Thiele, 1925

Gracilancilla lindae new species (Plate 25, Figures 3, 4)

Material Examined: HOLOTYPE - Length 16 mm, width 5 mm, trawled by commercial shrimpers from 35 m depth off Punto Fijo, Paraguana Peninsula, Gulf of Venezuela, Venezuela, 1979, USNM 859863.

Description: Shell very thin and fragile, very elongated with extremely protracted spire; shell without shoulder, fusiform; aperture elongated, narrow; anterior one-fourth of shell with separate enameled area; protoconch extremely large in proportion to shell size, rounded, dome-like; color pale cream-white, highly polished; columella with 2 plications; interior of aperture pale cream-white.

Etymology: Named for my wife, Linda J. Petuch.

Discussion: This amazing little species is only the second-known species of the previously-monotypic genus *Gracilancilla*. The type species of that genus, and the only other known *Gracilancilla*, is *G. sumatrana* (Thiele) from deep water off Indonesia. To find a second species of an Indonesian genus on the other side of the world, off Venezuela, is truly an exciting, if not problematical, biogeographical discovery.

Gracilancilla lindae differs from *G. sumatrana* in being a larger species with a larger aperture, by having a much larger protoconch, and in having a more inflated body whorl. In *Gracilancilla sumatrana*, the upper three whorls are brownish in color, while in *G. lindae* the entire shell is uniformly cream-white in color. At present, *G. lindae* is known only from the holotype. The poorly-known *G. sumatrana* has recently been illustrated by KAICHER (1982, card 3441) in her card-pack series on the Olividae.

Genus *Turrancilla* Martens, 1904

Turrancilla williamsoni new species (Plate 25, Figures 1, 2)

Material Examined: HOLOTYPE - Length 35 mm, trawled by commercial shrimpers from 35 m depth off Punto Fijo, Gulf of Venezuela, Venezuela, 1979, USNM 784573.

Description: Shell very slender and elongated, shiny and polished, with an elongated, protracted spire; fascicular band very small, restricted to anterior tip of shell; suture distinctly keeled; shell pale yellow-orange with darker orange enamel along shoulder and on spire whorls; fasciole darker orange; protoconch large, rounded, dome-like; interior of aperture pale orange; columella with 1 large, twisted plication; operculum unknown.

Etymology: Named for Dr. Peter Williamson, theoretical paleomalacologist, of the Museum of Comparative Zoology, Harvard University.

Discussion: *Turrancilla williamsoni* is the first member of the predominantly Japonic genus to be collected in the western Atlantic. The new species is closest to *T. suavis* (Yokoyama), but differs in being a slightly more inflated species, in being colored yellow-orange instead of tans and browns, and in having bright orange enamel on the spire whorls. The spire whorls of *Turrancilla williamsoni* are less enameled than those of *T. suavis*, and the keeled suture is more pronounced.

Subfamily **Olivinae**
Genus *Oliva* Bruguière, 1798
Subgenus *Strephona* Mörch, 1852

Oliva (Strephona) sargenti new species (Plate 17, Figures 2, 3)

Material Examined: HOLOTYPE - Length 29 mm, width 14 mm, 5 m depth on sand bottom, night diving with SCUBA, Malmok, Aruba Is., Netherlands Antilles, 1983, USNM 859864; PARATYPES - 2 specimens, same locality and depth as holotype, Petuch collection.

Description: Shell small for subgenus, ovate in outline, with rounded sides; spire low; body inflated, wider at mid-body than at shoulder; color pale bluish-white overlaid with loose network of very fine, pale tan triangles in netted pattern; triangles often aligned in zig-zag fashion, running length of shell from suture to anterior end; zig-zag and netted pattern overlaid with 2 wide bands of dark brown spots; one above mid-body, one below; in mature specimens, color pattern obscured by milky bluish-white enamel glaze; columella with 14-15 well-developed plications; anterior end of columella with characteristic reddish patch; edge of suture bordered with numerous fine, hairstreak brown flammules and intermittent large pale blue flammules; interior of aperture white; inner edge of lip bordered with dark brown flammules; protoconch large, pale tan.

Etymology: Named for Dennis M. Sargent, co-author of *Atlas of the Living Olive Shells of the World.*

Discussion: In the *Atlas of the Living Olive Shells of the World* (PETUCH and SARGENT, 1986, plate 30, figures 22, 23), I had originally referred this new species to the taxon *O. vermiculata* Gray. Since the publication of the book, I have received specimens of an olive from Haiti that closely resembles Gray's type of *O. vermiculata.* The Haitian olive is simply a zig-zag-patterned variant of *O. reticularis* Lamarck (the protoconchs and columellar plications are identical), so it appears that Gray's *vermiculata* should be relegated to the synonymy of *O. reticularis* (the form *vermiculata* Gray is shown here on Plate 17, Figure 10). The Aruban olive, then, is left without a name. Interestingly enough, the new species, *O. sargenti,* coexists with three other olive species; *O. fulgurator* Röding, which is endemic to the islands of Aruba, Curaçao, and Bonaire, and the coral atolls off Venezuela (*i.e.* Los Roques); *O. reclusa* Marrat, which is only found on Aruba, Curaçao, and Bonaire; and *O. reticularis,* which ranges from the Bahamas and Cuba to Barbados. *Oliva sargenti* differs from the Aruban-Bonairean population of *O. reticularis* (Plate 17, Figure 5) in being a more inflated shell with a lower spire, in having more numerous, well-developed columellar plications, in having a reddish patch on the anterior end of the columella, and in having a color pattern of a pale bluish-white base color with a fine netted pattern and two bands of brown dots as opposed to a white shell with large red triangles. *Oliva sargenti* coexists with the other Netherlands Antilles endemic, *O. reclusa* (Plate 17, Figure 4), but differs in having a much lower spire with a more rounded shoulder, in having more numerous and better-developed columellar plications, in having a much smaller protoconch, and in having the reddish patch on the anterior end of the columella. As far as is presently known, *O. sargenti* is restricted to the Malmok coast of Aruba, where it is relatively rare. In having an inflated, rounded shape and two bands of large brown spots, *O. sargenti* appears to be more closely related to the *O. tisiphona* Duclos species complex of Colombia and Venezuela than to the widespread Caribbean *O. reticularis.*

Family **Mitridae**

Genus *Subcancilla* Olsson and Harbison, 1953

Subcancilla leonardhilli new species (Plate 23, Figures 9, 10, 11)

Material Examined: HOLOTYPE - Length 18 mm, trawled by commercial shrimper from 35 m depth in Golfo de Triste, off Puerto Cabello, Venezuela, 1979, USNM 784574; PARATYPES - lengths 15 mm, 17 mm, same locality and depth as holotype, USNM.

Description: Shell elongated, fusiform; body whorl with 12 large, raised, evenly-spaced cords; 1 fine spiral thread between each pair of cords; whorls rounded; spire whorls with 3 large cords; columella with 3 large plications; color pale yellow with wide, pale brown axial flammules; interior of aperture pale yellow-white.

Etymology: Named for Mr. Leonard Hill of Miami, Florida, an avid student of the Mitridae.

Discussion: *Subcancilla leonardhilli* and the following species are the first-known Caribbean *Subcancilla* species. *Subcancilla leonardhilli* is closest to the Brazilian *S. lopesi* (Matthews and Coelho), but differs in being a more elongated shell with a much higher spire, by having more cords on the body whorl, and by being a more colorful shell, with reddish-brown axial flammules. The cords of *S. lopesi* are often brown in color, while those of *S. leonardhilli* are pale yellow. As far as is known, *S. leonardhilli* has only been collected in the Golfo de Triste, Venezuela, but probably extends around into the Gulf of Venezuela. I had originally referred to this new species as *S. illacidata* (Woodring), a Pliocene fossil from Jamaica (PETUCH, 1981), but now feel it is different enough to warrant a new name.

Subcancilla lindae new species (Plate 23, Figure 8)

Material Examined: HOLOTYPE - Length 24 mm, trawled by commercial shrimper from 35 m depth in Golfo de Triste, off Puerto Cabello, Venezuela, 1979, USNM 784575; PARATYPE - 1 specimen, USNM.

Description: Shell very narrow, elongated, with high, elevated spire; body whorl with 16 large, rounded, evenly-spaced spiral cords; suture impressed, producing tabulate whorls and stepped spire; sides of body whorl straight; aperture narrow; columella with 3 plications; color pure white; interior of aperture pure white.

Etymology: Named for my wife, Linda J. Petuch.

Discussion: I originally referred this new species to the taxon *Subcancilla venezuelana* (Hodson), a Pliocene fossil from Falcon, Venezuela (Petuch, 1981), but now feel it represents an unnamed but closely-related species. *Subcancilla lindae* is closest to the sympatric *S. leonardhilli* Petuch, but differs in being a more slender, elongated shell with finer and more numerous cords on the body whorl and spire, in having tabulate whorls and a stepped spire, and by being a pure white shell instead of yellow with reddish-brown flammules. Like the previous species, *S. lindae* has only been collected in the Golfo de Triste, Venezuela. *Subcancilla lindae* is similar, although smaller, to *S. gigantea* (Reeve) from the Panamic Province.

Family **Costellariidae**
Genus *Nodicostellaria* new genus

Diagnosis: Shells medium-sized for family, elongated, with rounded whorls; some species with angled shoulders; spires high, elevated; shells characteristically sculptured with close-packed and equal-sized spiral and axial ribs; intersection of spiral and axial ribs produce bead, altogether giving shell a uniformly pebbled appearance; inside of lip smooth, without ribbing; columella with 4 large plications that extend beyond the aperture and around siphonal canal; protoconch mamillate, usually with 2 smooth whorls; color pattern usually composed of broad bands and flammules of brown.

Type Species: *Costellaria kaicherae* Petuch, 1979, Abrolhos Archipelago, Brazil, (Plate 16, Figure 6).

Species in *Nodicostellaria: Nodicostellaria kaicherae* (Petuch), Abrolhos Archipelago, Brazil; *Nodicostellaria lixa* (Petuch), Abrolhos Archipelago, Brazil; *Nodicostellaria kremerae* Petuch, new species, described herein, Gulf of Venezuela.

Etymology: A combination of *Nodus* (Lt., "bump" or "lump") and *Costellaria,* the type genus of the family Costellariidae. In reference to the characteristic beaded sculpture.

Discussion: Within the family Costellariidae, the genus *Nodicostellaria* is closest to the genus *Uromitra* Bellardi, 1887, but differs in containing species with larger shells, species that lack the ribs inside the lip, and in containing species that lack the large *Pusia*-like axial ribs that are characteristic of *Uromitra.* In having fusiform shells with equal-sized and evenly-distributed surface beading, members of *Nodicostellaria* resemble small specimens of the mitrid genus *Scabricola.* The form of the protoconch, shape and form of the columellar plications, and general shell size, however, point to the inclusion of *Nodicostellaria* in the Costellariidae. *Vexillum styria* Dall and *Costellaria laterculatum* (Sowerby) may belong to this genus.

Nodicostellaria kremerae new species (Plate 16, Figure 13)

Material Examined: HOLOTYPE - Length 18 mm, width 5 mm, trawled by commercial shrimpers from 35 m depth off Puerto Fijo, Paraguana Peninsula, Gulf of Venezuela, Venezuela, 1797, USNM 859868.

Description: General shell form and sculpture pattern as for genus; shoulder angled; spiral and axial cords large, with intersections producing uniform beaded surface; color pale tan with darker tan cloudings; interior of aperture pale tan.

Etymology: Named for Mrs. Janet Kremer of Crystal Lake, Illinois, an enthusiastic student of the Costellariidae and Mitridae.

Discussion: *Nodicostellaria kremerae* is most similar to *N. lixa* (Petuch) from the Abrolhos Archipelago of Brazil, but differs in being a more elongated shell with a higher spire and more slender body whorl, and by having finer beaded sculpturing. *Nodicostellaria lixa* is a white shell with a pink anterior half, while *N. kremerae* is a tan shell with darker tan patches and axial streaks. Both species have a sharply angled shoulder.

Both *Nodicostellaria,* and the following new genus, *Turricostellaria,* appear to be restricted to the western Atlantic. No known species of either group have been found along the eastern Pacific or Panamic Province. These genera, then, may represent endemic western Atlantic species radiations.

Genus *Turricostellaria* new genus

Diagnosis: Shells very elongated, terebriform; suture indented, giving shell scalariform, turreted appearance; shells characteristically have impressed spiral line below and parallel to suture, producing wide subsutural band; shells ornamented with numerous axial ribs that run from suture to siphonal canal; axial ribs intersect subsutural band to produce row of elongated beads; siphonal canals short, stubby, but well-developed; columella with 4 strong plications; plications extend across columella and onto and around siphonal canal, producing 4 large cords.

Type Species: *Turricostellaria lindae* n.sp., described herein, 35 m depth off Cabo La Vela, Goajira Peninsula, Colombia.

Species In *Turricostellaria: Turricostellaria lindae* n.sp., northern Colombia; *Turricostellaria leonardhilli* n.sp., Gulf of Venezuela.

Etymology: A combination of *Turris* (Lt., "tower"), and *Costellaria,* the type genus of the family. In reference to the tabulate whorls and scalariform appearance that are characteristic of this genus.

Turricostellaria leonardhilli new species (Plate 16, Figure 12)

Material Examined: HOLOTYPE - Length 17 mm, width 5 mm, trawled by commercial shrimpers from 35 m depth off Punto Fijo, Paraguana Peninsula, Gulf of Venezuela, Venezuela, 1978, USNM 859869.

Description: General shell form as for genus; shell shiny, polished, without spiral threads; color white with orange band around mid-body; shoulder rounded; axial ribs low, rounded.

Etymology: Named for Mr. Leonard Hill of Miami, Florida, an ardent student of the Costellariidae.

Discussion: *Turricostellaria leonardhilli* differs from *T. lindae,* the only other species in the genus, in having a smooth shell without spiral sulci, in having rounded shoulders on the whorls, and in having an orange mid-body band. *Turricostellaria leonardhilli* is apparently a rare shell, having only been taken in the Gulf of Venezuela.

Turricostellaria lindae new species (Plate 21, Figures 14, 15)

Material Examined: HOLOTYPE - Length 17 mm, width 6 mm, trawled by commercial shrimpers from 35 m depth off Cabo La Vela, Goajira Peninsula, Colombia, 1976, USNM 859870; PARATYPE - 1 specimen, same locality and depth as holotype, Kevan Sunderland collection.

Description: General shell form as for genus; suture very incised, with squared-off subsutural band producing extreme tabulate appearance; spire scalariform; whorls with narrow, well-defined axial ribs; axial ribs overlaid, on body whorl, with 6 spiral impressed sulci and numerous fine spiral threads; color pale grayish-white.

Etymology: Named for my wife, Linda J. Petuch.

Discussion: *Turricostellaria lindae* differs from *T. leonardhilli,* its only known relative, in having sharply-defined tabulate whorls and a scalariform spire, in having better-developed axial ribs, in having impressed spiral sulci, and in lacking the orange mid-body band. At present, the new species is only known from the Goajira Peninsula area of Colombia.

Family **Volutomitridae**

Genus *Conomitra* Conrad, 1865

Conomitra leonardhilli new species (Plate 23, Figure 6, Plate 26, Figure 12)

Material Examined: HOLOTYPE - Length 15 mm, trawled by commercial shrimpers from 35 m depth in Golfo de Triste, off Puerto Cabello, Venezuela, 1979, USNM 784576; PARATYPES - 10 specimens, 12-18 mm lengths, same locality and depth as holotype, USNM; 2 adult specimens, same locality and depth as holotype, Leonard Hill collection.

Description: Shell fusiform, biconic; body with numerous fine axial ribs, running from suture to siphonal canal; axial ribs intersected by numerous spiral cords, giving shell reticulated sculpture; protoconch large, bulbous, composed of 2 whorls; aperture narrow; columella with 4 large plications; shell color white or pale cream-tan with numerous, reddish-brown zig-zag flammules; protoconch and interior of aperture white.

Etymology: Named for Mr. Leonard Hill of Miami, Florida.

Discussion: The genus *Conomitra* was thought to have been extinct since the late Miocene, and the discovery of *C. leonardhilli* and the following species shows that this prehistoric group has survived into the Recent. I originally identified this living species as *Conomitra lehneri* Jung, a Miocene fossil from Carriacou Is., Grenadines, Lesser Antilles (PETUCH, 1981, pp. 330-331), but now feel that it represents an un-named, descendant species. *Conomitra leonardhilli* is closest to *C. caribbeana* Weisbord (Plate 24, Figure 3), but differs in being a more slender species with a higher spire, in having finer axial and spiral sculpturing, and in having a zig-zag flammule color pattern instead of a banded one. *Conomitra leonardhilli* appears to be endemic to the Golfo de Triste, Venezuela. At present, there are three known living *Conomitra* species in the Caribbean: *C. leonardhilli* Petuch, Golfo de Triste and Cumana area of Venezuela, *C. lindae* Petuch, from off the coast of the Goajira Peninsula, Colombia, and *C. caribbeana* Weisbord, from the Gulf of Venezuela.

Conomitra lindae new species (Plate 24, Figures 1, 2)

Material Examined: HOLOTYPE - Length 10 mm, trawled by commercial shrimper from 35 m depth off Cabo La Vela, Goajira Peninsula, Colombia, 1974, USNM 784577.

Description: Shell fusiform, biconic, shiny and polished; body with numerous fine axial ribs intersected by numerous spiral ribs; small bead produced at intersection of axial spiral ribs; protoconch large, bulbous, composed of 2 whorls; columella with 4 large plications; color pale tan with 2 bands of arrow-shaped dark brown blotches; shoulder with intermittent dark brown patches; entire color pattern, in turn, overlaid with fine spiral bands of tiny, pale tan dots; protoconch brown; interior of aperture tan with 2 dark brown bands.

Etymology: Named for my wife, Linda J. Petuch.

Discussion: *Conomitra lindae,* third of the known living *Conomitra* species, is most similar to *C. caribbeana* from Venezuela, but differs in being a smaller, stockier shell with a more tabulate spire, in having a finer axial and spiral sculpture pattern, and in being a much more colorful shell, with large brown patches on the spire, shoulder, and body whorl. As far as is presently known, *C. lindae* is restricted to the upwelling systems off the Goajira Peninsula of Colombia, but may extend farther southward. *Conomitra lindae* is also the smallest of the three known species.

Family **Cancellariidae**

Genus *Aphera* H. and A. Adams, 1854

Aphera lindae new species (Plate 13, Figure 11)

Material Examined: HOLOTYPE - Length 10 mm, trawled by commercial shrimpers from 35 m depth in Golfo de Triste, off Puerto Cabello, Venezuela, 1974, USNM 784467.

Description: Shell solid, thickened, oval-elliptical in shape, shiny; entire shell surface covered with numerous fine axial ribs and numerous fine spiral cords; axial ribs and spiral cords of equal size, intersecting to produce finely cancellate sculpture; columella wide, reflected onto body whorl as polished, enameled area; columella with 2 large, equal-sized plications; edge of lip crenulated; interior of aperture with 15 lirae which disappear farther inside; color pure white with single large pale tan patch on dorsum; interior of aperture and columella white.

Etymology: Named for my wife, Linda J. Petuch.

Discussion: As in the genus *Conomitra,* the genus *Aphera* was though to have been extinct in the Caribbean, in this case, since the Pliocene. Unlike *Conomitra,* which was thought to be completely extinct, *Aphera* was known to have survived into the Recent as a single species from the Panamic Province of the eastern Pacific, *Aphera tessellata* (Sowerby). I originally referred the living Caribbean species to the taxon *A. islacolonis* (Maury), a Pliocene fossil from the Dominican Republic, Panama, and Costa Rica (PETUCH, 1981, p. 333), but now feel that it represents an un-named species that derived from the Pliocene *A. islacolonis. Aphera lindae* differs from *A. islacolonis* in being a smaller shell with a much finer cancellate-beaded sculpturing. The new species differs from its living Panamic relative, *A. tessellata,* by having a lower spire, by having a coarser sculpturing, and by being a white shell instead of dark brown banded and spotted one. *Aphera lindae* is known only from the Golfo de Triste, Venezuela.

Family **Conidae**

Genus *Conus* Linnaeus, 1758

Conus aureopunctatus new species (Plate 25, Figures 5, 6)

Material Examined: HOLOTYPE - Length 17 mm, width 8 mm, dredged by commercial shrimpers from 35 m depth in Gulf of Venezuela, off Punto Fijo, Falcon, Venezuela, 1977, USNM 859874.

Description: Shell small for genus, turnip-shaped, with wide body whorl and prominent constriction around anterior one-third, producing distinct anterior canal; shell shiny, polished; shoulder sharply carinated, with bladelike carina; spire elevated, scalariform; anterior third of shell ornamented with 10 thick, raised spiral cords, each separated from others by deeply-incised sulci; base color of shell white; smooth portion of body whorl with 4 rows of pale yellow-orange dots; spiral cords on anterior end marked with yellow-orange dots; spire whorls smooth, with numerous crescent-shaped orange flammules; interior of aperture white; periostracum brown, thick, and smooth.

Etymology: The name reflects the rows of yellow-orange dots that are characteristic of this new species.

Discussion: This distinctive little species is similar to only two other western Atlantic cones, *C. sennottorum* Rehder and Abbott from the Gulf of Mexico and *C. gibsonsmithorum* Petuch from Venezuela (PETUCH,1986) (Plate 25, Figure 7). From the former, *C. aureopunctatus* differs in being a smaller, more slender species with stronger, more regular cords on the anterior end. From the sympatric *C. gibsonsmithorum,* the new species differs in being a more slender shell, with a higher, more protracted spire, and by having rows of evenly-spaced, yellow-orange spots. The anterior spiral cords of *C. aureopunctatus* are also larger and better-developed than those of *C. gibsonsmithorum.* This new species appears to be endemic to the Gulf of Venezuela area.

Conus finkli new species (Plate 25, Figures 10, 11)

Material Examined: HOLOTYPE - Length 29 mm, width 14 mm, dredged by commercial shrimpers from 35 m depth in Gulf of Venezuela, off Punto Fijo, Falcon, Venezuela, 1979, USNM 859879; PARATYPES - 5 specimens, University of Miami, RSMAS Invertebrate Museum.

Description: Shell elongated, slightly turnip-shaped in outline; spire elevated, slightly concave on sides; body whorl ornamented with 25 evenly-spaced raised cords; cords on anterior canal with fine threads in-between; spire whorls ornamented with one large spiral cord, medially placed; color pure white with only one band of widely-spaced, reddish-brown crescent-shaped flammules around mid-body; spire whorls with widely-scattered, large, reddish-brown flammules; interior of aperture white; periostracum thick, dark brown, with rows of small hairs and with rows of large tufts around shoulder.

Etymology: Named for Dr. Charles Finkl, Jr., director of the Coastal Education and Research Foundation (CERF), Charlottesville, Virginia.

Discussion: *Conus finkli* appears to be closest to the more shallow water *C. cingulatus* Lamarck in shape, but differs in being more slender and having a higher spire. The spiral cords on *C. finkli* are also larger and more regular in arrangement than those of *C. cingulatus.* The wide-ranging northern South American *C. cingulatus* is also a more colorful shell, with vertical flammules and blotches of dark brown on a tan or purple shell and with a purple aperture. *Conus finkli* is virtually a colorless shell, in many ways approaching members of the *C. atractus* Tomlin species complex, such as *C. atractus austini* Rehder and Abbott from the Gulf of Mexico, *C. pseudoaustini* Usticke from Barbados, and *C. capricorni* Tursch, Van Mol, and Kemp from Surinam and Brazil. As far as is known, *C. finkli* is confined to the Gulf of Venezuela, in upwelling areas of cold water.

Conus kevani new species (Plate 25, Figures 8, 9)

Material Examined: HOLOTYPE - Length 17 mm, width 9 mm, dredged by commercial shrimpers from 35 m depth at the mouth of the Gulf of Venezuela, near the Monges Islands, Venezuela, 1979, USNM 859884; PARATYPES 2 specimens in the RSMAS collection, University of Miami.

Description: Shell small for genus, thin, fragile; spire elevated with slightly concave sides; shoulder strongly keeled; shoulder keel ornamented with prominent, beadlike coronations; coronations become stronger on last whorl; body whorl heavily sculptured with 30 large cords; fine spiral threads between cords; spire whorls ornamented with 3 thin spiral threads; color pure white with 2 broken bands of pale brown dots, one above mid-body, one below mid-body; interior of aperture white; periostracum thin, pale brown, smooth.

Etymology: Named for Mr. Kevan Sunderland of Ft. Lauderdale, Florida.

Discussion: Of all the small, sculptured cones found in the southern Caribbean, *C. kevani* is one of the only species that has a coronated spire. This one characteristic, alone, sets it apart from *C. cingulatus* Lamarck, *C. finkli* Petuch, and members of the *C. atractus* Tomlin complex. *Conus kevani* is also similar to *C. bajanensis* Usticke from Barbados, but differs in being a smaller species, in having a much flatter spire, in lacking the beading on the body whorl, and in being white instead of tan. *Conus kevani* is apparently restricted to the northern coast of South America.

Conus paraguana new species (Plate 25, Figures 12, 13)

Material Examined: HOLOTYPE - Length 18 mm, width 8 mm, dredged by commercial shrimpers from 35 m depth off Los Taques, Paraguana Peninsula, Falcon, Venezuela, Gulf of Venezuela,

1979, USNM 859889.

Description: Shell small for genus, thin, slender, smooth and polished; spire elevated, straight-sided; shoulder slightly rounded; anterior end with 8 deeply-impressed sulci; spire whorls smooth; color pale yellow with two wide bands of pale tan, closely-packed lines of dots and dashes; wide bands of dots separated by wide, white band around mid-body; white mid-body band with 2 lines of tiny tan dots; spire whorls heavily marked with numerous, large, yellow-tan flammules; interior of aperture white; periostracum thin, yellow, transparent, smooth.

Etymology: Named for the Paraguana Peninsula of Venezuela, the type locality.

Discussion: This distinctive little species doesn't resemble any known western Atlantic cone shell. Juveniles of *C. cingulatus* Lamarck could be confused with *C. paraguana,* but can easily be separated from the new species by being heavily sculptured, by having lower spires, and by being more darkly colored. In being a highly polished shell with bands composed of rows of dots, *C. paraguana* most closely resembles *C. dispar* Sowerby from the eastern Pacific Panamic Province, and may be its Atlantic analogue.

Conus parascalaris new species (Plate 25, Figures 14, 15)

Material Examined: HOLOTYPE - Length 23 mm, width 10 mm, dredged by commercial shrimpers from 35 m depth in Gulf of Venezuela, off Punto Fijo, Falcon, Venezuela, 1979, USNM 859890.

Description: Shell thin, fragile, elongated; spire elevated, protracted, scalariform; shoulder sharply angled; body whorl smooth, polished; anterior end with numerous fine, incised sulci; color white, overlaid with closely-packed, thin, vertical, pale tan flammules; spire with scattered tan, crescent-shaped flammules; interior of aperture white; periostracum thin, pale tan, with tufts along shoulder.

Etymology: In reference to the new species' resemblance to *Conus scalaris* of the Panamic Province.

Discussion: Like *Conus paraguana, C. parascalaris* most closely resembles a species that is not from the western Atlantic but is, instead, from the Panamic Province. The widespread eastern Pacific species *C. scalaris* Valenciennes is the closest relative of this new Venezuelan species. In the Caribbean region, *C. parascalaris* could only be confused with *C. paraguana,* but differs in having a much more protracted, scalariform spire, and by lacking the rows of dots which are characteristic of *C. paraguana.* From *C. scalaris, C. parascalaris* differs in being a much smaller, more fragile shell, and by having a color pattern composed of fine, vertical, hairline flammules. *Conus parascalaris* appears to be the Atlantic analogue of *C. scalaris.*

Conus perprotractus new species (Plate 25, Figures 16, 17)

Material Examined: HOLOTYPE - Length 46 mm, width 19 mm, dredged by commercial shrimpers from 35 m depth in Gulf of Venezuela, off Punto Fijo, Falcon, Venezuela, 1979, USNM 859891; PARATYPES - 2 specimens in RSMAS collection, University of Miami; 1 specimen, Robert Pace collection.

Description: Shell very elongated, tapering to a narrow anterior canal; spire low, flattened; shoulder angle sharp, slightly carinated; body whorl smooth, polished, with numerous faint striations around anterior end; spire whorls ornamented with 2 large spiral threads; color white overlaid with a complex color pattern comprising 2 thick orange-tan bands, one above mid-body and one below mid-body; posteriormost tan band contains row of darker orange checkers along margin of mid-body, and scattered amorphous flammules along margin of shoulder; anteriormost tan band continuous, regular in form; mid-body marked with wide, pure white band; anterior third and anterior tip of shell lavender purple; spire whorls with scattered orange-tan flammules; interior of aperture white; periostracum unknown.

Etymology: In reference to the characteristic extreme elongation of shell form.

Discussion: This unusual new species is closest to *Conus carioca* Petuch from Brazil (Plate 16, Figure 18), but differs in being a much more elongated shell with a lower spire, and by lacking the bright orange and pink color bands that are so characteristic of *C. carioca.* The bright lavender-purple anterior end of *C. perprotractus* sets it aside from *C. carioca* and all other related forms. *Conus villepini* form *fosteri* Clench from Bermuda, is also similar to *C. perprotractus,* especially in having a low spire, but differs in being less colorful, less elongated, thinner and more lightweight, and in being less polished. The purple tip and bright orange-tan color bands of *C. perprotractus* easily separate the new species from any morph of *C. villepini.* Although the holotype and paratypes were dead when collected, and are faded and drilled by *Cliona* sponges, the color pattern is still well-preserved. A live-taken specimen would have dark orange or tan bands and a vivid lavender-purple anterior end.

Conus tristensis new species (Plate 16, Figures 9, 10, 11)

Material Examined: HOLOTYPE - Length 37 mm, width 29 mm, trawled by commercial shrimp boats from 35 m depth off Tucacas, Carabobo State, Golfo de Triste, Venezuela, 1979, USNM 784470.

Description: Shell stocky with compressed body, wide across shoulder and tapering rapidly toward anterior end; shoulder sharply angled, slightly carinated; spire low, flattened; body whorl sculptured with 18 prominent, raised spiral cords; spiral cords pustulated; spire whorls sculptured with six incised spiral sulci; shell color pure white with small, scattered pale orange-brown flammules; spire whorls with regularly-spaced, amorphous brown flammules; early whorls pale orange; periostracum thick, with rows of erect hairs that correspond to raised, pustulated cords on body whorl.

Etymology: Named for the Golfo (Gulf) de Triste, Venezuela, the type locality.

Discussion: *Conus tristensis* is very similar to the widespread Pliocene Caribbean *C. symmetricus* Sowerby, and probably represents the last living member of the *C. symmetricus-dominguensis* species complex. I originally (PETUCH, 1981) referred this Venezuelan cone to *C. symmetricus,* but now feel that it represents a different, although closely-related, species. *Conus tristensis* differs from the Pliocene *C. symmetricus* in being a more tapered, elongated shell, with fewer cords on the body whorl, and with finer, less-developed sculpturing on the spire whorls. Like *C. symmetricus,* the cords on the body whorl are pustulated, giving a rough texture to the shell. *Conus tristensis* resembles no other living Caribbean species.

Conus venezuelanus new species (Plate 25, Figures 18, 19)

Material Examined: HOLOTYPE - Length 27 mm, trawled by commercial shrimpers from 25 m depth off Puerto Cabello, Golfo de Triste, Venezuela, 1979, USNM 784469; PARATYPES - length 21 mm, 25 mm, same depth, locality, and date as holotype, USNM.

Description: Shell elongated, tapered, with sharp-angled shoulder; spire elevated, concave along sides; body whorl heavily sculptured with numerous fine spiral threads and sulci; spire whorls sculptured with 3 spiral cords; color white to pale salmon-pink, with 2 bands of dark yellow maculations around mid-body; spire marked with scattered small, brown, crescent-shaped flammules; interior of aperture white; periostracum thin, smooth, translucent yellow.

Etymology: Named for Venezuela, country of the type locality.

Discussion: As in *Conus tristensis,* I originally referred this species to a Pliocene fossil, *C. planiliratus* Sowerby (PETUCH, 1981). *Conus venezuelanus,* however, differs from *C. planiliratus* in having much finer sculpturing on the body whorl and by having more numerous spiral cords. The only Recent cone that bears any resemblance to *C. venezuelanus* is *C. stimpsoni* Dall, from the Carolinian Province areas off the Carolinas, Georgia, Florida, and the Gulf of Mexico. *Conus venezuelanus* differs from *C. stimpsoni* in being a consistently more slender shell, by having two bands of yellow maculations around the mid-body, by having a heavily sculptured spire, and by having numerous incised spiral sulci on the body whorl.

Conus bayeri new species (Plate 17, Figures 15, 16)

Material Examined: HOLOTYPE - Length 16 mm,trawled by commercial shrimper from 35 m depth in Golfo de Morrosquillo, Colombia, 1975, USNM 859875.

Description: Shell small for genus, slender with low spire; body whorl smooth, with 10 incised sulci around anterior tip; shoulder sharp-angled, carinated; spire whorls slightly canaliculate; 2 small spiral sulci on body whorl just below shoulder carina; shell color white, with 6 wide spiral bands made up of small, pale yellow-tan vertical lines; clear band around mid-body; anterior tip white; spire whorls with scattered pale yellow-tan flammules; interior of aperture white; early whorls with beaded carina.

Etymology: Named for Dr. Frederick M. Bayer, Dept. of Invertebrate Zoology, National Museum of Natural History, Smithsonian Institution.

Discussion: *Conus bayeri* most closely resembles the Venezuelan *C. paraguana* Petuch, but differs in having a lower, canaliculate spire, in having two sulci around the shoulder below the carina, and in having bands of tiny vertical lines instead of rows of dots. *Conus bayeri* appears to be closer to members of the *C. gradatus* species complex from the Panamic Province-Eastern Pacific than it does to other Caribbean cones. A relatively non-descript little species, *C. bayeri* is apparently rare and has only been collected in the Golfo de Morrosquillo of Colombia.

Conus colombianus new species (Plate 17, Figures 11, 12)

Material Examined: HOLOTYPE - Length 22 mm, trawled by commercial shrimper from 35 m depth off Islas del Rosario, Colombia, 1977, USNM 859877.

Description: Shell small for genus, stocky, broad across shoulder; spire low, flattened; shoulder sharp-angled; body whorl smooth, with 10 small spiral cords around anterior end; spire with 4 spiral threads; shell pale yellow with 4 closely-spaced brown lines around body whorl just below (anterior of) mid-body; brown flammules and white blotches run through 4 lines and extend over anterior tip; body whorl above (posterior of) mid-body line without markings or pattern; spire marked with large, evenly-spaced orange-tan flammules; spire flammules extend onto sharp edge of shoulder, giving shoulder checkered appearance; interior of aperture white.

Etymology: Named for Colombia, the country of the type locality.

Discussion: Although dead-collected and quite faded, enough characteristics remain to show that *Conus colombianus* is quite different from any other known Caribbean cone shell. In shape, and in having a smooth, sharp-angled shoulder, it somewhat resembles *C. mayaguensis* Usticke from Puerto Rico (endemic to that island), but differs in having a two-toned color pattern, with an unpatterned posterior half and a flammuled and lined anterior half. *Conus colombianus* appears to belong to the *C. magellanicus* Hwass species complex and is the only species of the group to have such a two-toned color pattern. A fresh specimen would probably be orange with darker orange-brown markings. This new species may be endemic to the coral reef areas around the archipelago of the Islas del Rosario.

Family **Turridae**
Subfamily **Borsoninae**
Genus *Paraborsonia* Pilsbry, 1922

Paraborsonia lindae new species (Plate 26, Figures 10, 11)

Material Examined: HOLOTYPE - Length 16 mm, trawled by commercial shrimper from 35 m depth in Golfo de Triste, off Puerto Cabello, Venezuela, 1979, USNM 784581; PARATYPES - 3 specimens, lengths 14-16 mm, same depth and location as holotype, USNM.

Description: Shell short, stocky, biconic in shape; shoulder rounded, demarcated by 2 large, beaded cords; body whorl ornamented with 20-25 fine, beaded spiral cords; subsutural area indented, slightly canaliculate; columella with 3 large plications; interior of aperture with numerous ribs; protconch large, flattened on top; anal notch corresponding to canaliculate subsutural area, deep and narrow; shell color pale yellow with darker yellow crescent-shaped flammules along the shoulder; anterior tip of columella dark yellow; interior of aperture white.

Etymology: Named for my wife, Linda J. Petuch.

Discussion: I had originally called this species *Paraborsonia varicosa* (Sowerby), which is a Pliocene fossil from the West Indies (PETUCH, 1981). Although similar in appearance to *P. varicosa,* the new Venezuelan living species differs in being a smaller shell with finer sculpturing, and in having a much more rounded shoulder without a distinct carina. I feel that these differences are significant enough to give the living species a name of its own. Like the genera *Aphera* and *Conomitra, Paraborsonia* was though to be extinct in the Atlantic Ocean since the Pliocene. *Paraborsonia lindae* further reinforces the "prehistoric," archaic nature of the molluscan fauna of Venezuela, and is one of the more important faunal components of the Gatunian Relict Pocket of northern South America.

Subfamily **Turrinae**
Genus *Polystira* Woodring, 1928

Polystira lindae new species (Plate 17, Figures 13, 14)

Material Examined: HOLOTYPE - Length 19 mm, width 6 mm, trawled from 35 m depth in the Gulf of Venezuela, off Punto Fijo, Paraguana Peninsula, Venezuela, 1978, USNM 859895; PARATYPES - 1 specimen, same locality and depth as holotype, Kevan Sunderland collection.

Description: Shell small for genus, slender, elongated; whorls rounded, without shoulder; body whorl with 5 large, sharp-edged cords; spire whorls with 3 cords; siphonal canal with numerous fine spiral threads; anal notch narrow, deep; shell color pure white; interior of aperture white; periostracum thick, velvety.

Etymology: Named for my wife, Linda J. Petuch.

Discussion: *Polystira lindae* is the smallest-known member of the genus. It is closest to *P. macra* Bartsch, from off Puerto Rico, but differs in being a much smaller, less elongated species, by having fewer cords on the body whorl, and by lacking the sharp shoulder angle of *P. macra. Polystira lindae* is only known from the Gulf of Venezuela.

Class **Bivalvia**
Subclass **Palaeotaxodonta**
Order **Nuculoida**
Superfamily **Nuculacea**
Family **Nuculanidae**

Genus *Scissuladrana* new genus

Diagnosis: Shells extremely fragile, translucent, spathate in shape, flattened; hinge line of anterior half of shell serrated, with row of tiny, raised tooth-like projections; anterior one-third and posterior two-thirds of shell surface separated by anteriorly-angled demarcation line; one-third of shell anterior to line smooth, glassy, with only very faint, microscopic concentric lines; two third of shell posterior to line heavily sculptured with coarse scissulations that run about 30-40 degrees to the underlying concentric lines; umbo elevated above hinge line; ligament internal, attached to bilobate chondrophore.

Type Species: *Scissuladrana ludmillae* Petuch, n.sp. (described herein), Recent of northern Colombia, by monotypy.

Etymology: In reference to the characteristic oblique-angled scissulate sculpturing.

Discussion: The new genus *Scissuladrana* differs from the closely-related genus *Adrana* H. and A. Adams in being broader, stubbier, with more blade-shaped shells (similar to *Yoldia* in shape), in having truncated anterior ends, and in having coarse, obliquely-angled scissulated sculpture on the posterior two-thirds of the shell, and by having a split sculpture pattern, with a smooth, glassy, and highly-polished anterior one-third and an obliquely-ribbed posterior two-thirds. The oblique scissulate rib pattern may be an adaptation for living in shallow, surfy areas, where the oblique ribs help keep the animal buried in the sand. *Scissuladrana,* then, may be a recent intertidal offshoot of the deeper, quieter-water genus *Adrana.*

Scissuladrana ludmillae new species (Plate 26, Figures 1, 2, 3: Text Figure 1)

Material Examined: HOLOTYPE - Length 40 mm, single valve, on beach (high tide line) north of Boquilla, Cartagena, Colombia, Jan. 1977, USNM859899; PARATYPE - same locality and date as holotype, USNM 859900.

Description: As for description of genus.

Etymology: Named for Mrs. Ludmilla Strickland of Mount Holly, New Jersey.

Discussion: At present the monotypic genus *Scissuladrana* and *S. ludmillae* are only known from the beach areas near Cartagena, Colombia.

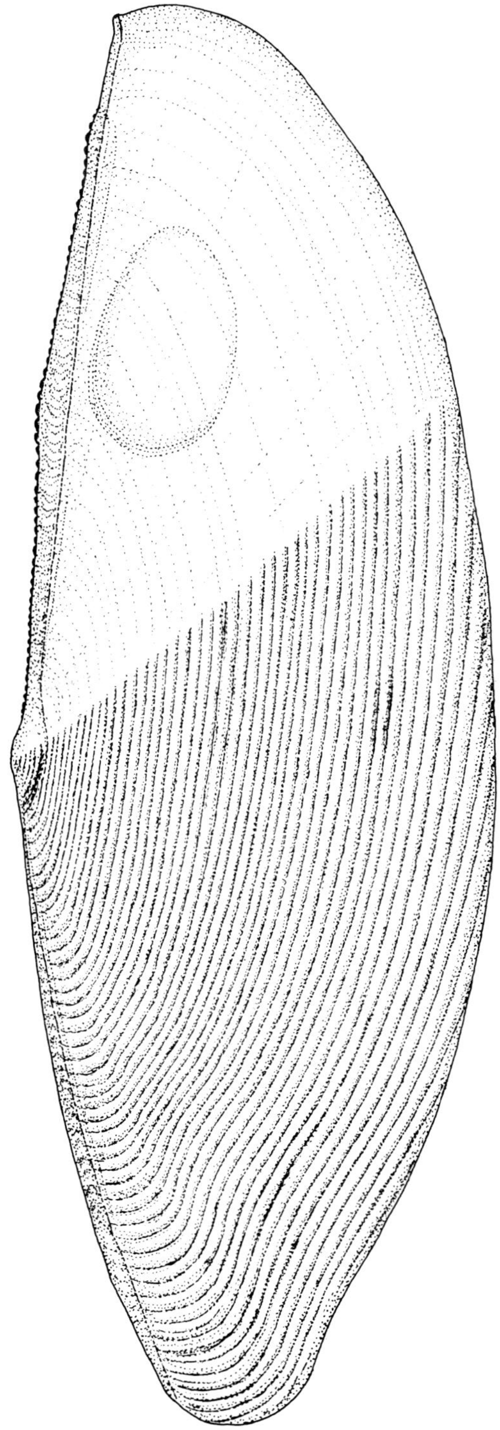

Text Figure 1. Right valve of paratype of *Scissuladrana ludmillae* Petuch, n.gen., n.sp., showing detail of oblique, scissulate sculpture.

PLATE 16. NEW AND SELDOM-SEEN GASTROPODS FROM THE WESTERN ATLANTIC

Figures 1, 2: *Sassia lewisi* Harasewych and Petuch, 1980, dorsal and ventral views of paratype, 14 mm, 140 m depth off St. James, Barbados.

Figure 3: *Sassia lewisi* Harasewych and Petuch, 1980, same specimen coated with ammonium hydroxide to enhance sculpture.

Figures 4, 5: *Murexiella iemanja* Petuch, 1979, dorsal and ventral views of holotype, 8 mm, Abrolhos Islands, Brazil.

Figure 6: *Nodicostellaria kaicherae* (Petuch, 1979), dorsal view of holotype, 8 mm, Abrolhos Islands, Brazil.

Figures 7, 8: *Oliva (Strephona) maya* Petuch and Sargent, 1986, dorsal and ventral views of 65 mm specimen, 30 m depth off Contoy Island, Yucatan, Mexico.

Figure 9: *Conus tristensis* Petuch, n.sp., view of spire, showing characteristic flammules.

Figures 10, 11: *Conus tristensis* Petuch, n.sp., dorsal and ventral views of holotype, 37 mm.

Figure 12: *Turricostellaria leonardhilli* Petuch, n. gen and n.sp., ventral aspect of holotype, 17 mm.

Figure 13: *Nodicostellaria kremerae* Petuch, n.gen. and n.sp., ventral aspect of holotype, 18 mm.

Figures 14, 15: *Conus penchaszadehi* Petuch, 1986, dorsal and ventral views of holotype, 14 mm, Goajira Peninsula, Colombia.

Figures 16, 17: *Conus riosi* Petuch, 1986, dorsal and ventral views of holotype, 18 mm, Salvador, Bahia State, Brazil.

Figure 18: *Conus carioca* Petuch, 1986, dorsal view of 57 mm specimen, Rio de Janeiro, Brazil.

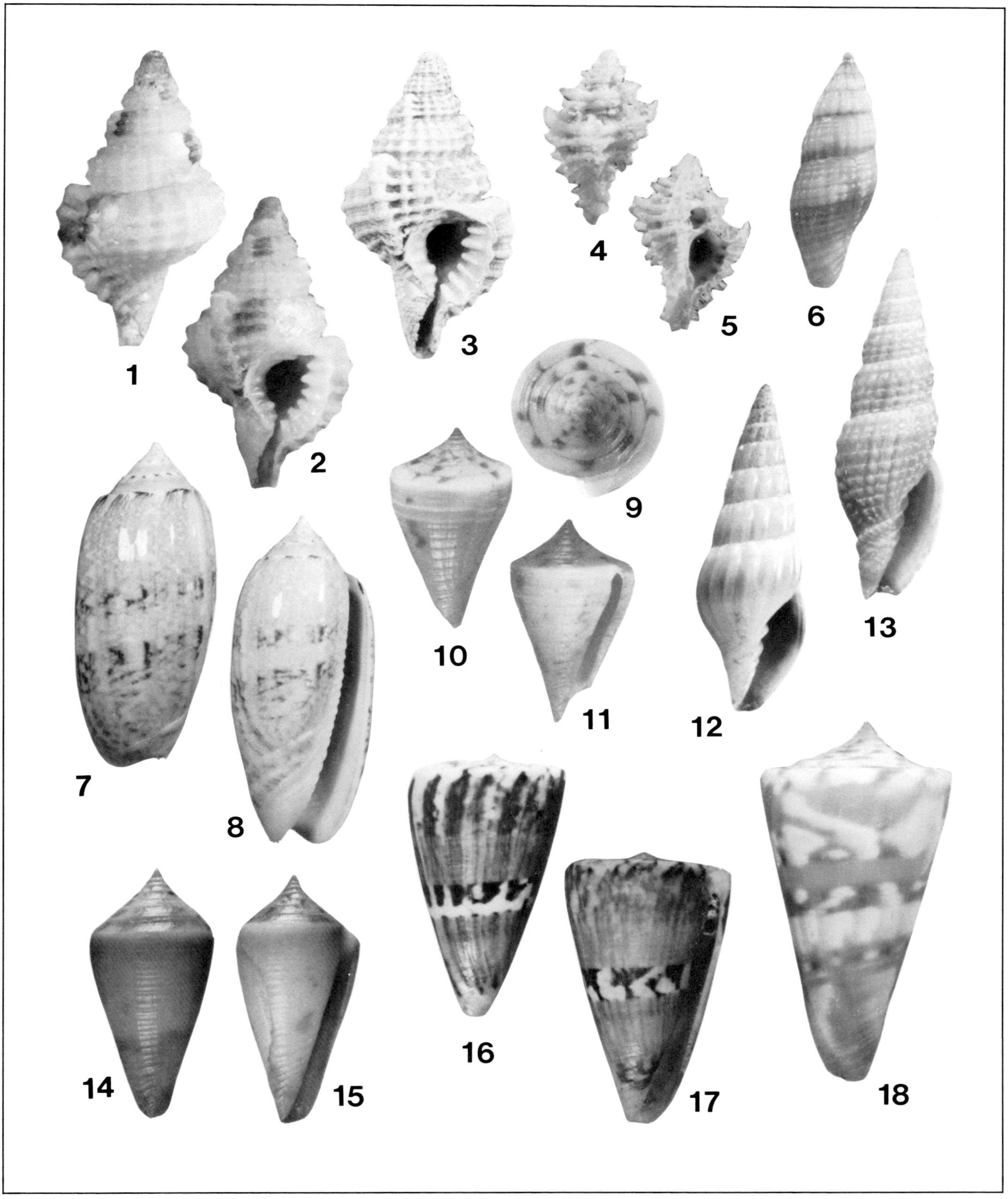
1
2
3
4
5
6
7
8
9
10
11
12
13
14
15
16
17
18

PLATE 17. NEW AND SELDOM-SEEN GASTROPODS FROM HONDURAS, NETHERLANDS ANTILLES AND SOUTH AMERICA

Figure 1: *Eburna balteata* (Swainson, 1825), ventral view of 25 mm specimen, Malmok, Aruba Is.

Figures 2, 3: *Oliva sargenti* Petuch, n.sp., dorsal and ventral views of holotype, 29 mm.

Figure 4: *Oliva reclusa* Marrat, 1871, dorsal view of 40 mm specimen, Malmok, Aruba Is.

Figure 5: *Oliva reticularis* Lamarck, 1811, dorsal view of 31 mm specimen, Lac Bay, Bonaire Is.

Figures 6, 7: *Murexiella hilli* Petuch, n.sp., dorsal and ventral views of holotype, 34 mm.

Figures 8, 9: *Muricopsis aliceae* Petuch, n.sp., dorsal and ventral views of holotype, 11 mm, Bonaire Island.

Figure 10: *Oliva reticularis* form *vermiculata* Gray, 1858, dorsal view of 26 mm specimen, Gonave Is., Haiti.

Figures 11, 12: *Conus colombianus* Petuch, n.sp., dorsal and ventral views of holotype, 22 mm.

Figures 13, 14: *Polystira lindae* Petuch, n.sp., dorsal and ventral views of holotype, 19 mm.

Figures 15, 16: *Conus bayeri* Petuch, n.sp., dorsal and ventral views of holotype, 16 mm.

Figures 17: *Conus sunderlandi* Petuch, n.sp., 38 mm specimen, pale color form, Utila Is., Honduras.

Figures 18, 19: *Lyria (Enaeta) reevei* (Dall, 1907),dorsal and ventral views of 25 mm specimen, Roatan Is., Honduras.

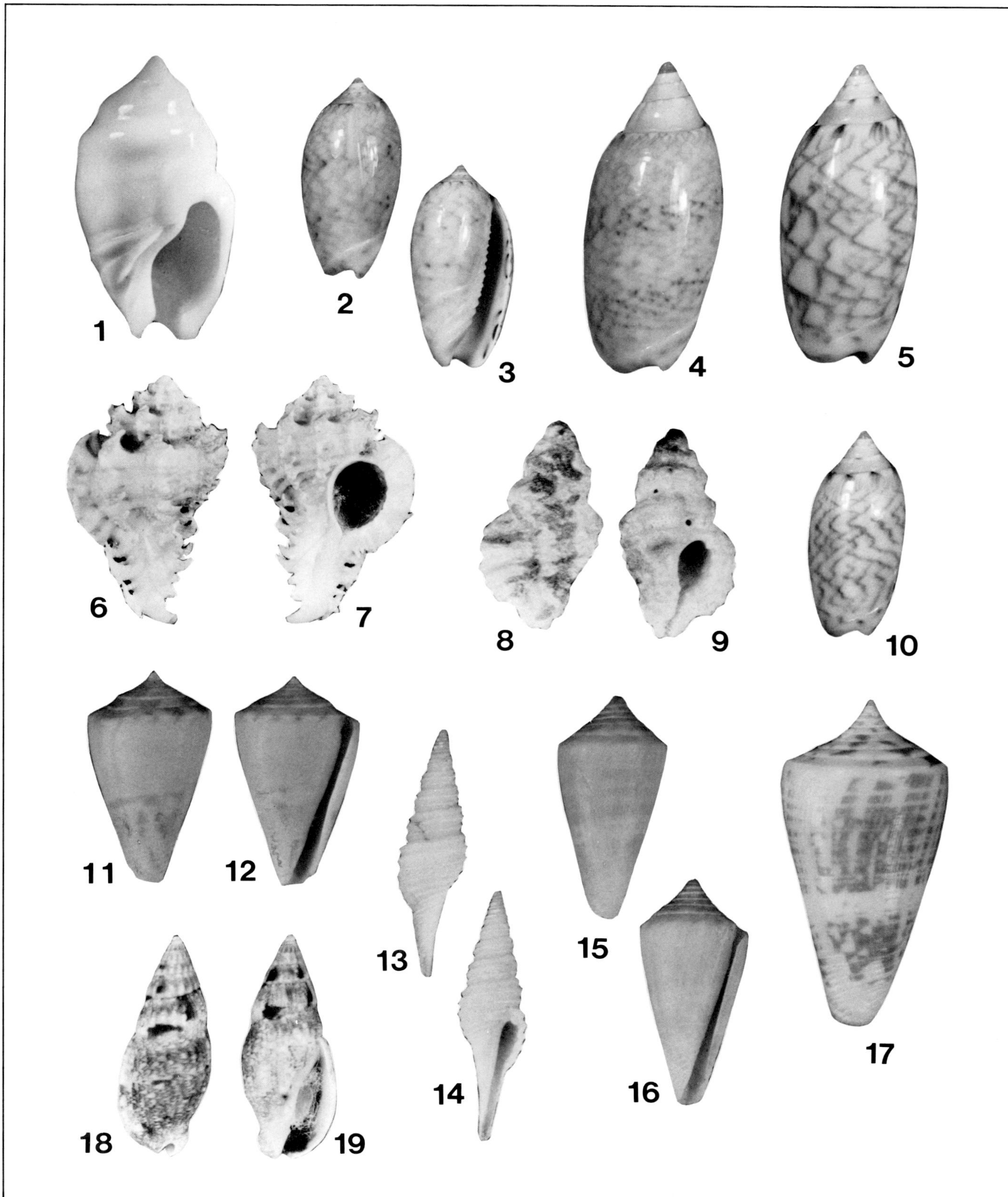
1
2
3
4
5
6
7
8
9
10
11
12
13
14
15
16
17
18
19

PLATE 18. CHARACTERISTIC GASTROPODS FROM NORTHERN SOUTH AMERICA

Figure 1: *Latirus bernadensis* Bullock, 1974, 100 m depth off Barbados, northern South America, and Brazil.

Figure 2: *Fulgurofusus brayi* (Clench, 1959), 250 m depth off Venezuela, 40 mm.

Figures 3, 4: *Fulgurofusus brayi* (Clench, 1959), 150 m depth off Golfo de Uraba, Colombia, low-spired variety, 37 mm.

Figure 5: *Murex donmoorei* Bullis, 1964, 35 m depth off Venezuela, 65 mm.

Figure 6: *Siratus springeri* (Bullis, 1964), holotype, Guyana.

Figure 7: *Voluta musica* Linnaeus, 1758, 90 mm specimen from Goajira Peninsula, Colombia.

Figure 8: *Fusiturricula jaquensis* (Sowerby, 1850), 35 m depth off Punto Fijo, Venezuela, 46 mm.

Figures 9, 10: *Siratus briskasii* (Verrill, 1953), 16 mm specimen from 35 m depth off Punto Fijo, Venezuela.

Figure 11: *Voluta virescens* Lightfoot, 1786, 43 mm specimen from Cartagena, Colombia, low tide on beach.

Figure 12: *Scaphella (Clenchina) evelina* Bayer, 1971, dark brown color form, 111 mm, from 500 m depth off Golfo de Morrosquillo, Colombia.

Figures 13, 14: *Strombina galba* Weisbord, 1962, 11 mm specimen from 35 m depth off Punto Fijo, Venezuela.

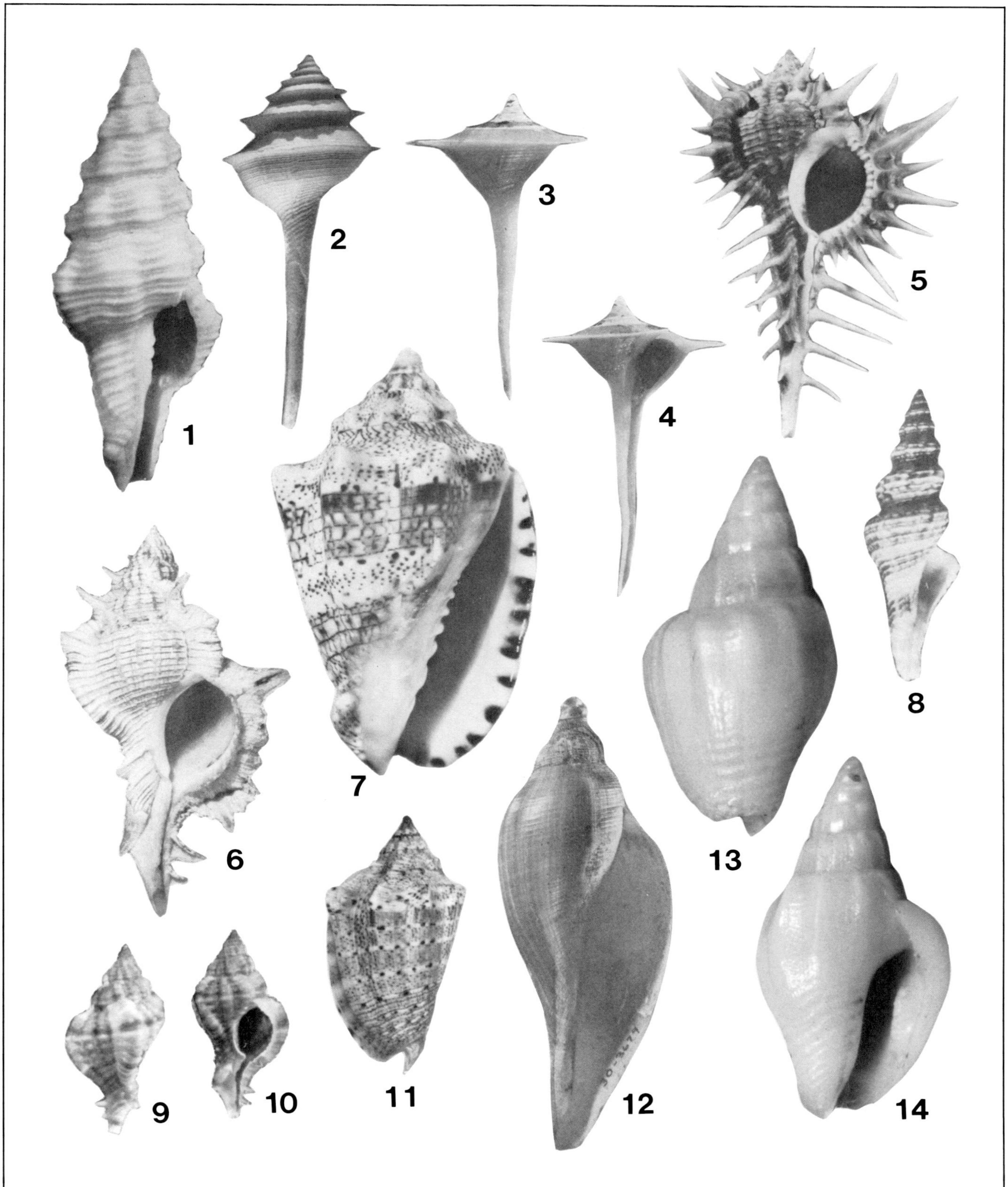
1
2
3
4
5
6
7
8
9
10
11
12
13
14

PLATE 19. CHARACTERISTIC GASTROPODS FROM NORTHERN COLOMBIA AND THE GULF OF VENEZUELA

Figures 1, 2: *Phyllonotus cf. globosus* (Emmons, 1858), 60 mm specimen from Amuay Bay, Gulf of Venezuela.

Figure 3: *Fusinus caboblanquensis* Weisbord, 1962, 140 mm specimen from Gulf of Venezuela, low tide.

Figures 4, 5: *Conus puncticulatus* Hwass, 1792, 18 mm specimen from Amuay Bay, Gulf of Venezuela, low tide.

Figure 6: *Conus puncticulatus* form *mauritianus* Hwass,1792, (dark color form), 21 mm, on beach, Cartagena, Colombia.

Figure 7: *Eburna glabrata* (Linnaeus, 1758), 75 mm specimen, 30 m depth off Goajira Peninsula, Colombia.

Figure 8: *Amalda venezuelana* (Weisbord, 1962), 32 mm specimen, 35 m depth off Punto Fijo, Venezuela.

Figure 9: *Amalda tankervillei* (Swainson, 1825), 52 mm specimen, 10 m depth off Isla Margarita, Venezuela.

Figures 10, 11: *Panamurex velero* E. Vokes, 1970, 12 mm specimen, 2 m depth on barnacles, Santa Marta, Colombia.

Figure 12: *Conus undatus* Kiener, 1845, 22 mm specimen from 35 m depth of Punto Fijo, Venezuela.

Figure 13: *Hindsiclava chazaliei* (Dautzenberg, 1900), 33 mm specimen, 35 m depth off Punto Fijo, Venezuela.

Figure 14: *Polystira barretti* (Guppy, 1866), 70 mm specimen with attached *Crucibulum marensis* Weisbord, 1962, Venezuela.

Figure 15: *Persicula multilineata* (Sowerby, 1846), holotype 13 mm, Venezuela.

Figure 16: *Persicula porcellana* (Gmelin, 1791), 14 mm, northern Colombia; common off Golfo de Morrosquillo.

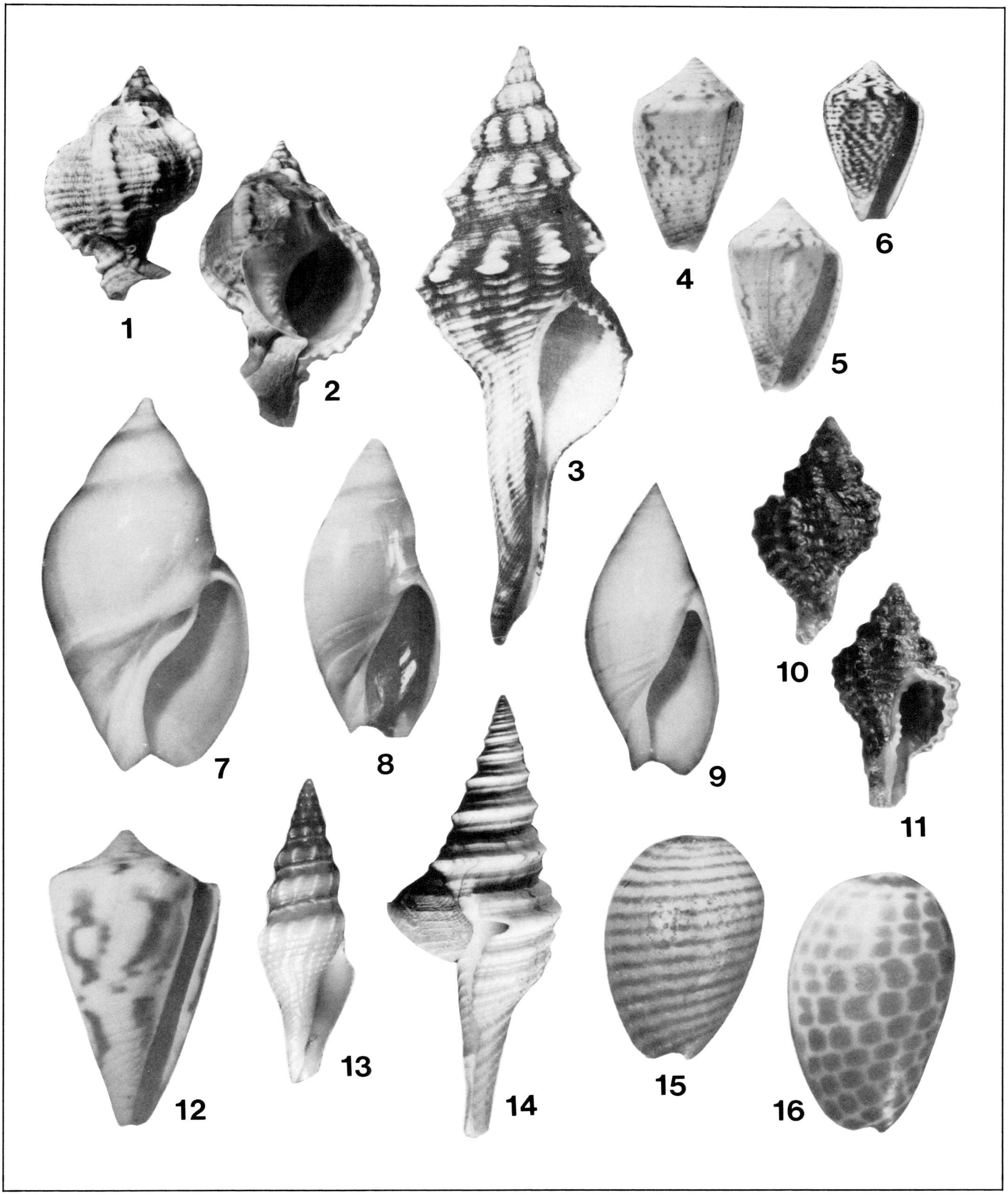
1
2
3
4
5
6
7
8
9
10
11
12
13
14
15
16

PLATE 20. CHARACTERISTIC GASTROPODS FROM NORTHERN COLOMBIA AND VENEZUELA

Figures 1, 2: *Cypraea (Siphocypraea) donmoorei* (Petuch, 1979), dorsal and ventral views of 58 mm specimen, Gulf of Venezuela.

Figure 3: *Siratus thompsoni* (Bullis, 1964), ventral view of holotype.

Figures 4, 5: *Panamurex recticanalis* (Weisbord, 1962), dorsal and ventral views of 18 mm specimen, Gulf of Venezuela.

Figure 6: *Fasciolaria tulipa hollisteri* Weisbord, 1962, dorsal view of 105 mm specimen, Gulf of Venezuela.

Figure 7: *Voluta musica guinaica* Lamarck, 1811, ventral view of 60 mm specimen, Amuay Bay, Gulf of Venezuela.

Figure 8: *Conus granarius* Kiener, 1845, dorsal view of 33 mm specimen, Goajira Peninsula, Colombia.

Figure 9: *Clathodrillia gibbosa* (Born, 1778), ventral view of 45 mm specimen, Gulf of Venezuela.

Figure 10: *Conus sanguineus* Kiener, 1845, ventral view of 23 mm specimen, off Los Taques, Gulf of Venezuela. Shell smooth with blood-red markings.

Figure 11: *Terebra petiveriana* Deshayes, 1857, dorsal view of 33 mm specimen, Cartagena, Colombia.

Figure 12: *Oliva tisiphona* form *oblonga* Marrat, 1870, 52 mm specimen, 3 m depth off Santa Marta, Colombia.

Figure 13: *Conus cf. granarius* Kiener, 1845, 26 mm specimen, 35 m depth off Golfo de Uraba, Colombia.

Figure 14: *Conus mappa caracanus:* Hwass, 1792, 65 mm specimen, 35 m depth in Golfo de Triste, Venezuela.

Figure 15: *Strombina caboblanquensis* Weisbord, 1962, 20 mm specimen from 35 m depth, Gulf of Venezuela.

Figure 16: *Strombina pumilio* (Reeve, 1859), 16 mm specimen, on beach, Riohacha, Goajira Peninsula, Colombia.

Figure 17: *Fusiturricula iole* Woodring, 1928, 17 mm specimen, 35 m depth off Golfo de Triste, Venezuela.

Figure 18: *Strioterebrum trispiralis* (Weisbord, 1962), 14 mm specimen, Adicora, Paraguana Peninsula, Venezuela.

Figure 19: *Strioterebrum quadrispiralis* (Weisbord, 1962), 13 mm specimen, Adicora, Paraguana Peninsula, Venezuela.

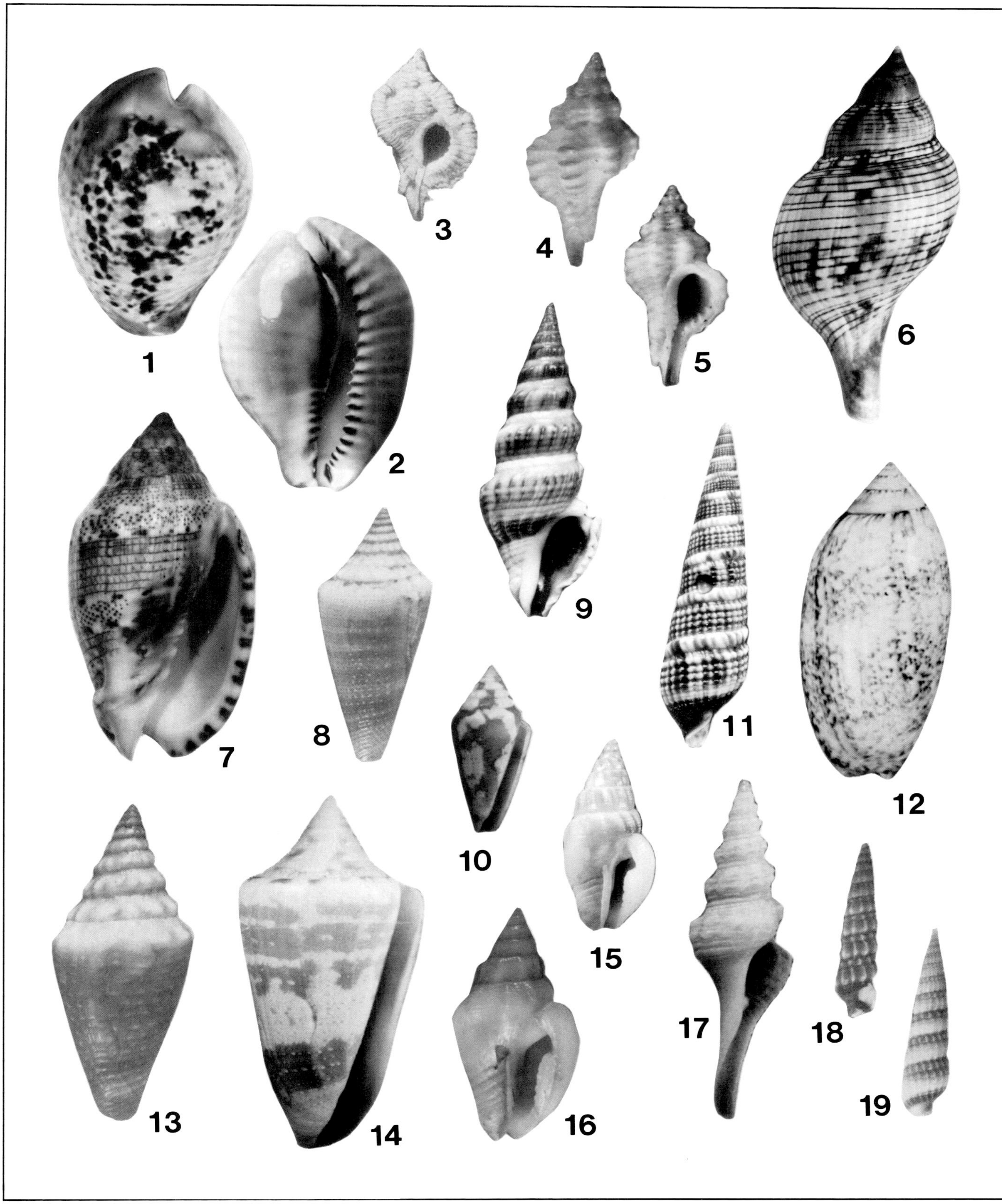
1
2
3
4
5
6
7
8
9
10
11
12
13
14
15
16
17
18
19

PLATE 21. NEW AND SELDOM-SEEN GASTROPODS FROM NORTHERN SOUTH AMERICA

Figure 1: *Scaphella (Clenchina) evelina* Bayer, 1971, 92 mm holotype, 500 m depth off Isla Fuerte, Colombia.

Figures 2, 3: *Fulgurofusus brayi* (Clench, 1959), 150 m depth off Cartagena, Colombia, 42 mm specimen.

Figure 4: *Conus hedgesi* Sowerby, 1913, 24 mm specimen, 35 m depth in Gulf of Venezuela.

Figures 5, 6: *Murex sunderlandi* Petuch, n.sp., dorsal and ventral views of holotype, 41 mm.

Figure 7: *Fulgurofusus brayi* (Clench, 1959), 34 mm specimen, 250 m depth off Colon, Panama, flat-spired form.

Figures 8, 9: *Mohnia kaicherae* Petuch, n.sp., dorsal and ventral views of holotype, 29 mm.

Figure 10: *Conus venezuelanus* Petuch, n.sp., 25 mm specimen, 35 m depth in Gulf of Venezuela.

Figures 11, 12: *Murexsul harasewychi* Petuch, n.sp., dorsal and ventral views of holotype, 15 mm.

Figure 13: *Siratus briskasii* (Verrill, 1953), dark color form, 14 mm, 35 m depth in Gulf of Venezuela.

Figures 14,15: *Turricostellaria lindae* Petuch, n.gen. and n.sp., dorsal and ventral views of holotype, 17 mm.

Figures 16, 17: *Truncaria lindae* Petuch, n.sp., dorsal and ventral views of holotype (fragmentary), 16 mm, with reconstructions of spire based on outline of *Truncaria filosa* (Adams and Reeve).

Figure 18: *Truncaria lindae* Petuch, n.sp., 13 mm fragment (paratype), Cabo Cordera, Venezuela.

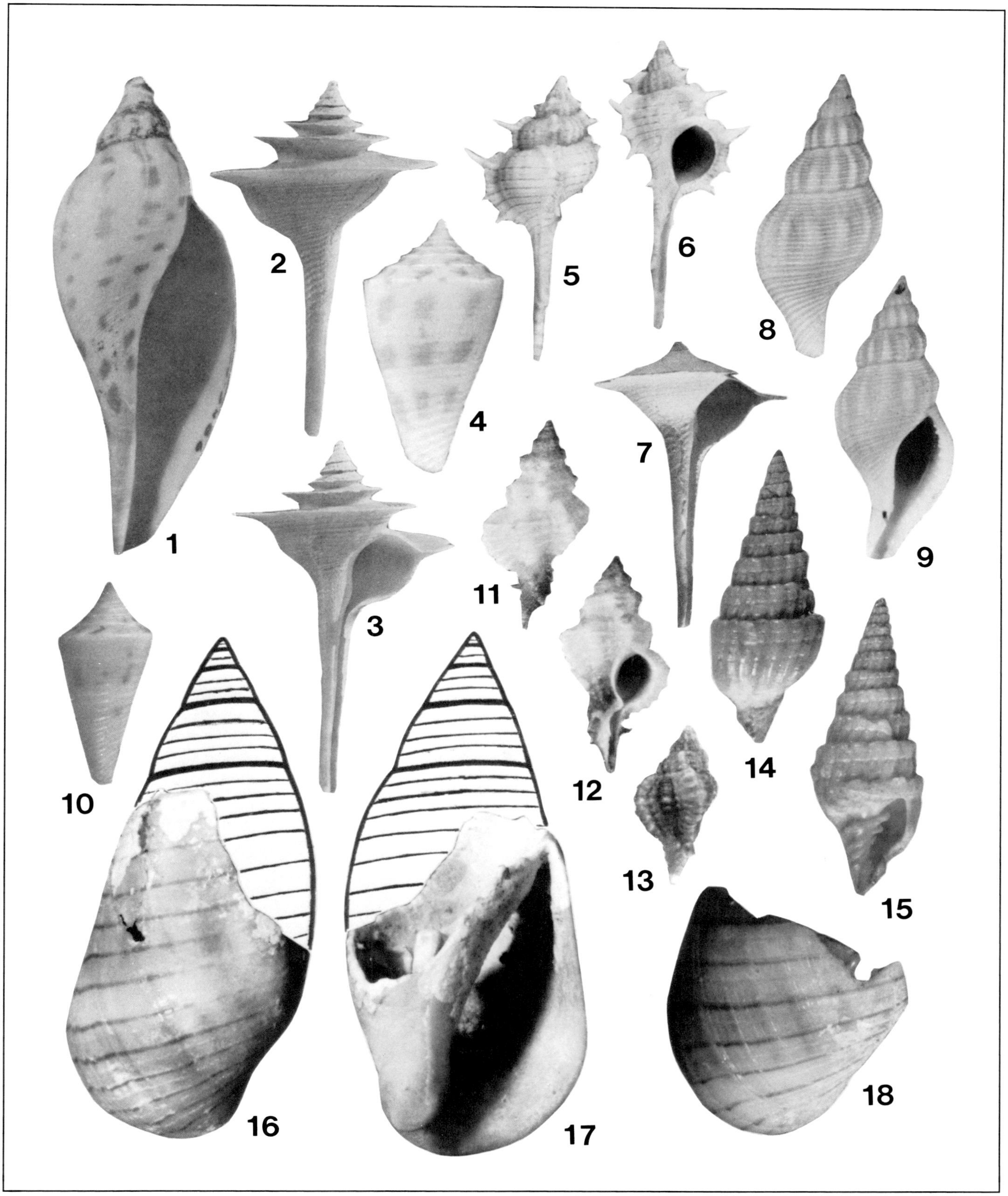
1
2
3
4
5
6
7
8
9
10
11
12
13
14
15
16
17
18

PLATE 22. COWRIES FROM NORTHERN SOUTH AMERICA AND COZUMEL ISLAND

Figures 1, 2: *Cypraea aubryana* Jousseaume, 1869, dorsal and ventral views of 45 mm specimen, 120 m depth off Cozumel Island.

Figures 3, 4: *Cypraea (Siphocypraea) mus mus* Linnaeus, 1758, dorsal and ventral views of 51 mm specimen, Amuay Bay, Paraguana Peninsula, Gulf of Venezuela.

Figures 5, 6: *Cypraea (Siphocypraea) mus tristensis* Petuch, n.subsp., dorsal and ventral views of holotype, 48 mm.

Figures 7, 8: *Cypraea (Siphocypraea) donmoorei* (Petuch, 1979), dorsal and ventral views of holotype, 64 mm, 37 m depth off Cabo La Vela, Goajira Peninsula, Colombia, USNM 770731.

Figures 9, 10: *Cypraea (Siphocypraea) mus tristensis* Petuch, n.subsp., dorsal and ventral views of paratype.

Figure 11, 12: *Cypraea (Siphocypraea) donmoorei* (Petuch, 1979), dorsal and ventral views of 52 mm specimen, 35 m depth off Goajira Peninsula, Colombia.

Figure 13: *Cypraea (Siphocypraea) donmoorei* (Petuch, 1979), juvenile bulla specimen, 55 mm.

Figure 14: *Cypraea (Siphocypraea) mus mus* Linnaeus, 1758, juvenile bulla specimen, 34 mm.

Figures 15, 16: *Discotrivia labiosa* (Gaskoin, 1836), dorsal and ventral views of 19 mm specimen, 100 m depth of Cabo La Vela, Goajira Peninsula, Colombia.

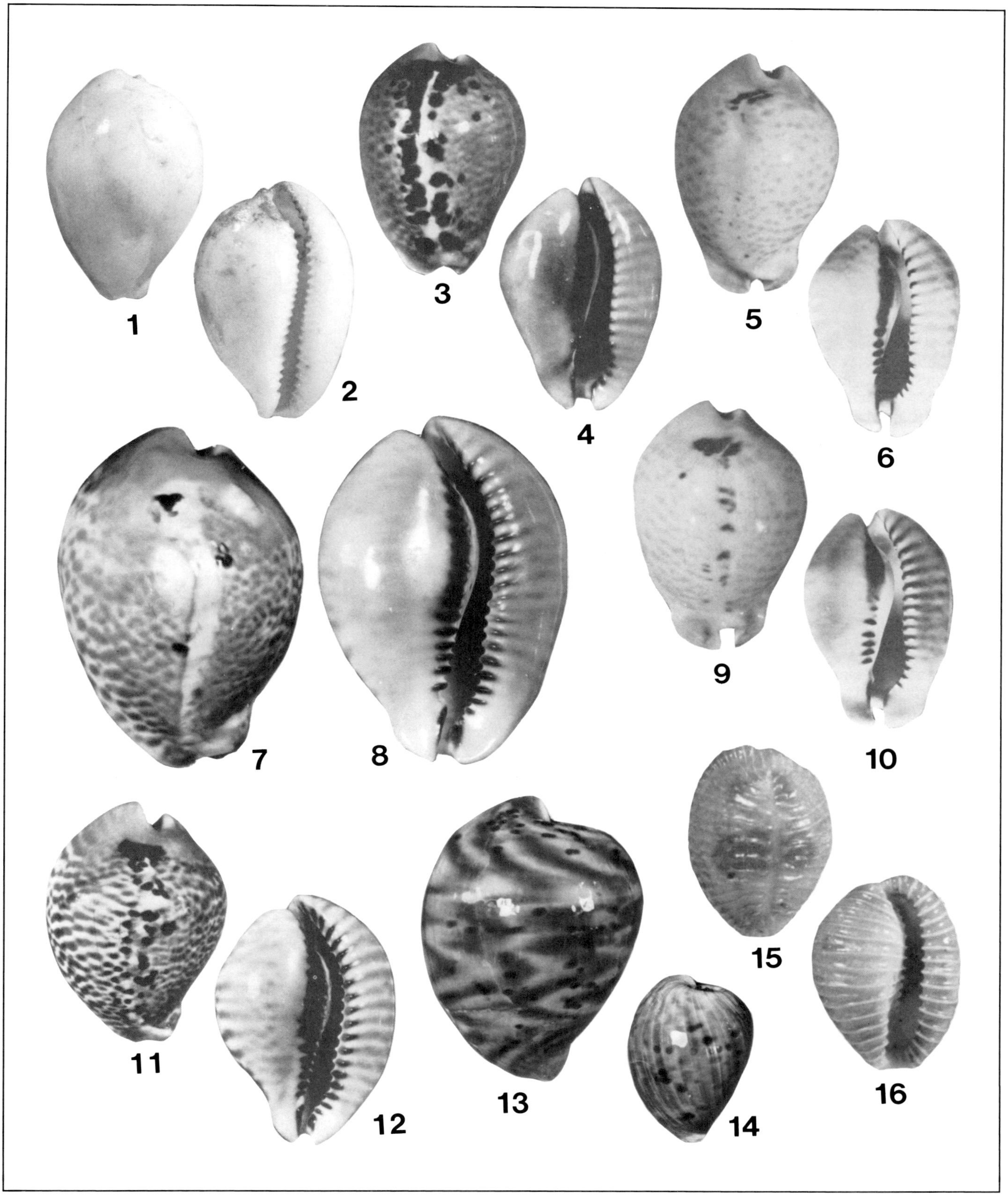

PLATE 23. NEW AND SELDOM-SEEN GASTROPODS FROM NORTHERN SOUTH AMERICA

Figures 1, 2: *Morum (Cancellomorum) lindae* Petuch, n.sp., dorsal and ventral views of holotype, 35 mm.

Figures 3, 4: *Morum (Cancellomorum) dennisoni* (Reeve, 1842), dorsal and ventral views of 51 mm specimen, off Arrowsmith Bank, Yucatan, Mexico (for comparison with *Morum lindae).*

Figure 5: *Sconsia nephele* Bayer, 1971, dorsal view of holotype, off Grenada, Lesser Antilles.

Figure 6: *Conomitra leonardhilli* Petuch, n.sp., dorsal and ventral view of holotype, 15 mm, showing detail of columellar plications.

Figure 7: *Volutomitra erebus* Bayer, 1971, ventral view of holotype, 36 mm, off Goajira Peninsula, Colombia.

Figure 8: *Subcancilla lindae* Petuch, n.sp., ventral view of holotype, 24 mm.

Figure 9: *Subcancilla leonardhilli* Petuch, n.sp., ventral view of paratype, 15 mm.

Figures 10, 11: *Subcancilla leonardhilli* Petuch, n.sp., dorsal and ventral views of holotype, 18 mm.

1
2
3
4
5
6
7
8
9
10
11

PLATE 24. NEW AND SELDOM-SEEN GASTROPODS FROM THE GOAJIRA PENINSULA AND VENEZUELA

Figures 1, 2: *Conomitra lindae* Petuch, n.sp., dorsal and ventral views of holotype, 10 mm.

Figure 3: *Conomitra caribbeana* Weisbord, 1929, ventral view of 12 mm specimen, Isla Margarita, Venezuela.

Figures 4, 5: *Antillophos bayeri* Petuch,n.sp.,dorsal and ventral views of holotype, 17 mm.

Figure 6: *Antillophos candei* (Orbigny, 1842), 27 mm specimen, off Cabo La Vela, Goajira Peninsula, Colombia.

Figures 7, 8: *Antillophos cf. adelus* (Schwengel, 1942), 23 mm specimen, Goajira Peninsula, Colombia.

Figures 9, 10: *Antillophos elegans* (Guppy, 1866), 20 mm specimen, 35 m depth in Golfo de Triste, Venezuela.

Figures 11, 12: *Colubraria bayeri* Petuch, n.sp., dorsal and ventral views of holotype, 17 mm.

Figures 13, 14: *Cryptospira martini* (Petit, 1853), 38 mm specimen, 35 m depth off Cumana, Venezuela.

Figures 15, 16: *Marginella marginata* Born 1778, form *crassa* Röding, 1798, 25 mm specimen from 20 m depth off Goajira Peninsula, Colombia.

Figures 17, 18: *Dermomurex (Trialatella) kaicherae* Petuch, n.sp., dorsal and ventral views of holotype, 164 mm.

PLATE 25. NEW GASTROPODS FROM VENEZUELA AND THE GULF OF VENEZUELA

Figures 1, 2: *Turrancilla williamsoni* Petuch, n.sp.,dorsal and ventral views of holotype, 35 mm.

Figures 3, 4: *Gracilancilla lindae* Petuch, n.sp., dorsal and ventral views of holotype, 16 mm.

Figures 5, 6: *Conus aureopunctatus* Petuch, n.sp., dorsal and ventral views of holotype, 17 mm.

Figure 7: *Conus gibsonsmithorum* Petuch, 1986, ventral view of paratype, 18 mm, Gulf of Venezuela.

Figures 8, 9: *Conus kevani* Petuch, n.sp., dorsal and ventral views of holotype, 17 mm.

Figures 10, 11: *Conus finkli* Petuch, n.sp., dorsal and ventral views of holotype, 29 mm.

Figures 12, 13: *Conus paraguana* Petuch, n.sp., dorsal and ventral views of holotype, 18 mm.

Figures 14, 15: *Conus parascalaris* Petuch, n.sp.,dorsal and ventral views of holotype, 23 mm.

Figures 16, 17: *Conus perprotractus* Petuch, n.sp., dorsal and ventral views of holotype, 46 mm.

Figures 18, 19: *Conus venezuelanus* Petuch, n.sp., dorsal and ventral views of holotype, 27 mm.

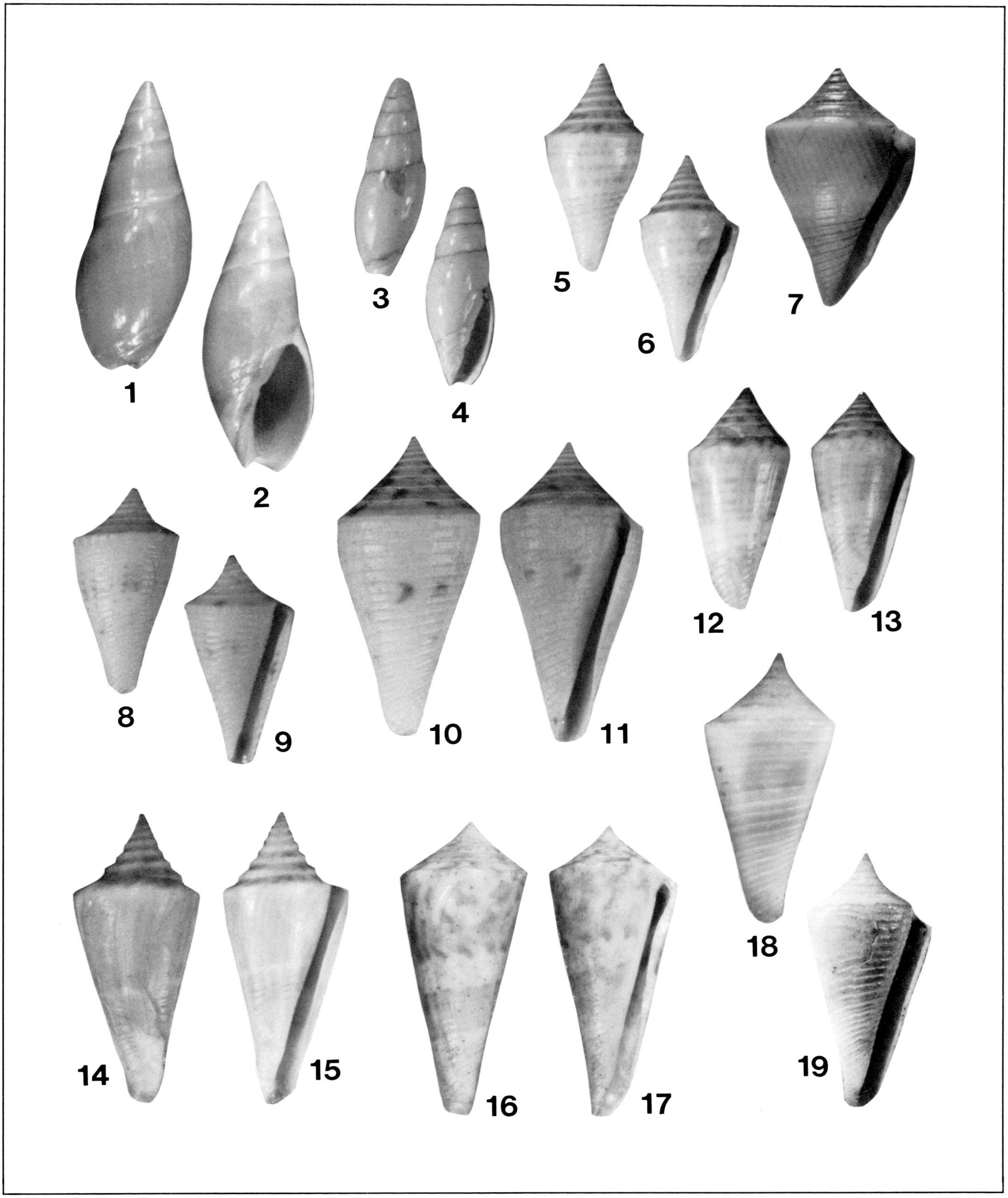
1
2
3
4
5
6
7
8
9
10
11
12
13
14
15
16
17
18
19

PLATE 26. NEW AND SELDOM-SEEN BIVALVES AND GASTROPODS FROM NORTHERN COLOMBIA AND VENEZUELA

Figures 1, 2: *Scissuladrana ludmillae* Petuch, n.gen., n.sp., exterior and interior views of holotype, 40 mm.

Figure 3: *Scissuladrana ludmillae* Petuch, n.gen., n.sp., paratype, Boquilla, Cartagena, Colombia.

Figure 4: *Adrana notabilis* Rehder, 1939, 53 mm specimen, Boquilla, Cartagena, Colombia.

Figure 5: *Adrana tellinoides* (Sowerby, 1823), 23 mm specimen, Boquilla, Cartagena, Colombia.

Figure 5A: *Adrana scaphoides* Rehder, 1939, 50 mm specimen, Boquilla, Cartagena, Colombia.

Figures 6, 7: *Pseudocyphoma gibsonsmithorum* Petuch, n.sp., dorsal and ventral views of holotype, 17 mm.

Figures 8, 9: *Phenacovolva (Turbovula) lindae* Petuch, n.sp., dorsal and ventral views of holotype, 17 mm.

Figures 10, 11: *Paraborsonia lindae* Petuch, n.sp., dorsal and ventral views of holotype, 16 mm.

Figure 12: *Conomitra leonardhilli* Petuch, n.sp., ventral view of holotype, 16 mm.

Figures 13, 14: *Homalopoma finkli* Petuch, n.sp., dorsal and ventral views of holotype, 18 mm.

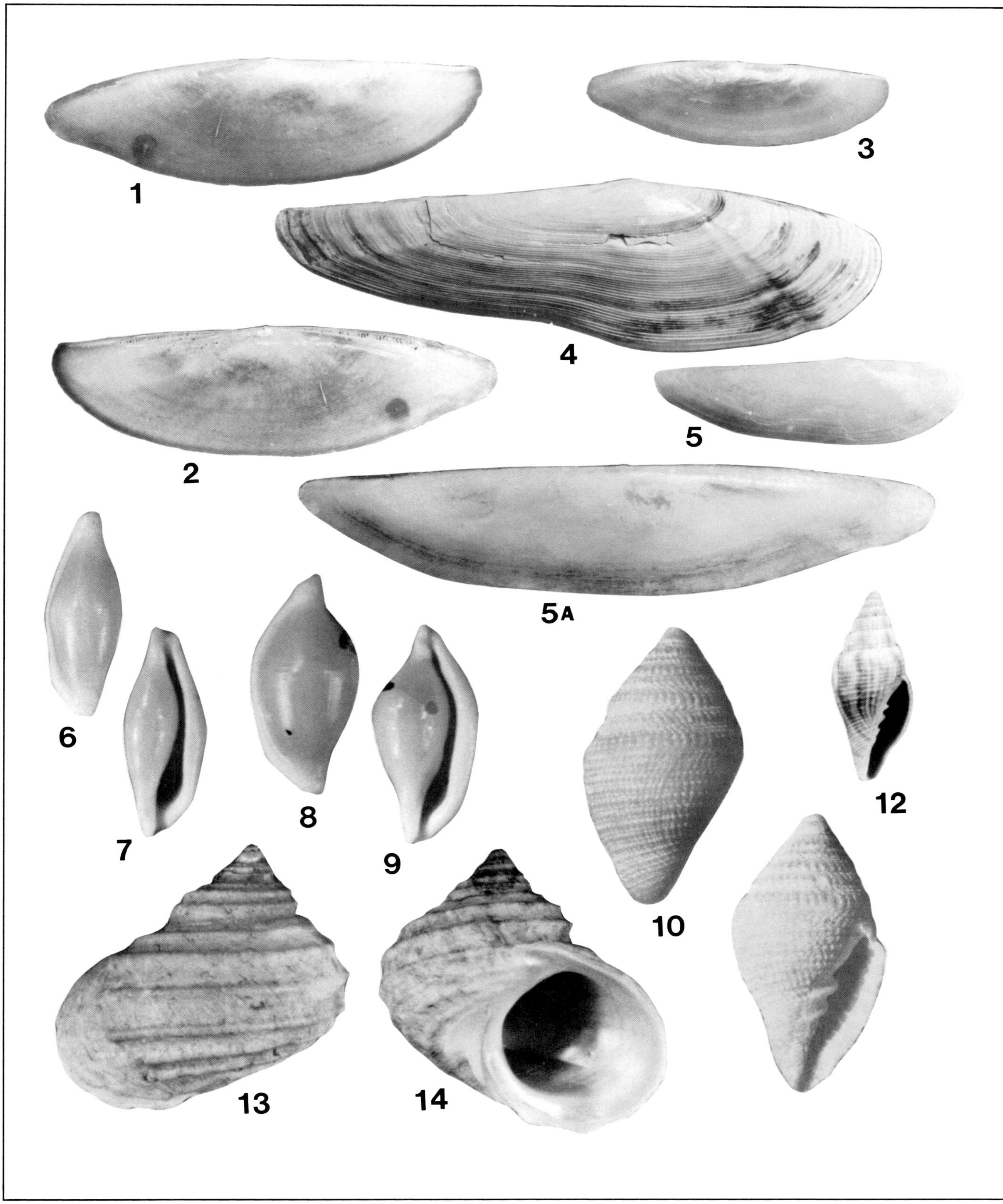
1
3
4
2
5
5A
6
7
8
9
10
12
13
14

CHAPTER 7

New Discoveries from Brazil

Although Brazil is technically outside of the Caribbean region, it contains many species that are of Caribbean origin. Along with many endemic species, the northern and central coasts of Brazil contain molluscan assemblages that include common Caribbean species such as *Cypraea zebra, Cypraea spurca acicularis, Cypraea cinerea, Cassis tuberosa, Strombus costatus,* and *Strombus pugilis.* In being related to the Caribbean fauna to the north, I felt it was important to discuss this rich and still poorly-known faunal region in this book. Two areas are of particular interest in the Brazilian region; the offshore regions of Pará, Pernambuco, Bahia, and Espiritu Santo States, and the huge coral reef complexes of the Abrolhos Archipelago off southern Bahia State.

The Abrolhos Archipelago and reef complex comprises five islands, two major coral reef complexes (one, the Parcel das Paredes being over 30 km on a side), and innumerable smaller, isolated reefs. Until my research trip to the Abrolhos Archipelago in 1977, the molluscan fauna of that region was completely unknown. While visiting the Archipelago and reef complex, an entire new western Atlantic molluscan fauna was discovered and officially reported on two years later (PETUCH, 1979a). Besides containing typical Brazilian endemic species such as *Strombus goliath* Schröter, *Vasum cassiforme* (Kiener), *Turbinella laevigata* Anton, *Oliva circinata* Marrat, and *Pleuroploca aurantiaca* (Lamarck), the Abrolhos Archipelago was host to a new muricid fauna comprising *Dermomurex (Trialatella) oxum* Petuch (Plate 27, Figures 10, 11), *Muricopsis oxossi* Petuch (Plate 15, Figures 15, 16), and *Murexiella iemanja* Petuch (Plate 16, Figures 4, 5), a new costellariid fauna including *Nodicostellaria kaicherae* (Petuch) (Plate 16, Figure 6) and *Nodicostellaria lixa* (Petuch) (Plate 27, Figure 16), a bizarre new volute *Lyria (Plicoliva) zelindae* (Petuch) (Plate 27, Figures 3, 4, 5), a new fasciolariid, *Latirus ogum* Petuch (Plate 27, Figures 1, 2), a new ovulid *Cyphoma macumba* Petuch (Plate 15, Figure 17, 18), and an unusual cone shell fauna comprising *Conus beddomei* Sowerby, *Conus iansa* Petuch (Plate 27, Figures 8, 9), and a new species which is described in this chapter. Future research work in the Abrolhos Archipelago will undoubtedly uncover many new endemic species.

The offshore areas of northern and central Brazil, particularly off Salvador, Bahia State, contain many new and un-named species. One of the most beautiful is a large *Murexiella* species (described in this chapter) that belongs to the *Murexiella macgintyi* species complex of the northern Caribbean. Other discoveries from off northern Brazil include the beautiful new turbinellid, *Fulgurofusus ecphoroides* Harasewych (Plate 27, Figures 12, 13), and a new fasciolariid that is endemic to Fernando de Noronha Island, *Latirus vermeiji* Petuch (Plate 27, Figures 6, 7). To the south of the Abrolhos Archipelago, along the coasts of Espiritu Santo and Rio de Janeiro States, fisherman trawling in 30-35 m depths have uncovered an entire new cone shell fauna, one of the most distinctive in the western Atlantic. Included within this endemic conid radiation are *Conus carioca* Petuch (Plate 28, Figures 15, 16; Plate 16, Figure 18), *Conus riosi* Petuch (which has recently been taken off the Lesser Antilles) (Plate 16, Figure 16, 17), *Conus clerii* Reeve (Plate 28, Figure 17), *Conus clenchi* Martins, *Conus tostesi* Petuch (Plate 27, Figures 14, 15), *Conus xanthocinctus* Petuch (Plate 28, Figures 13, 14), *Conus cf. latifasciatus* Sowerby (Plate 28, Figures 9, 10), and *Conus pusillus* Lamarck (Plate 28, Figures 11, 12).

New Species From Brazil

Murexiella leonardhilli new species
Oliva circinata tostesi new subspecies
Conus abrolhosensis new species

Family **Muricidae**
Genus *Murexiella* Clench and Farfante, 1943

Murexiella leonardhilli new species (Plate 28, Figures 5, 6)

Material Examined: HOLOTYPE - Length 31 mm, trawled from 10 m depth in Baia de Todos os Santos, off Salvador, Bahia, Brazil, USNM 859843; PARATYPE - 1 specimen, same locality as holotype, Robert Work collection.

Description: Shell large for subgenus with compressed, flattened body whorl; shoulder sharply-angled, very wide in proportion to shell; siphonal canal long, straight; 6 thick varices per whorl; body whorl with 6 thick cords between varices; short, stubby, recurved small spine on varix where intersected by cord; siphonal canal with 3 spines; no secondary spines between 3 siphonal spines; aperture oval-circular; shell color pinkish-orange with darker subsutural band.

Etymology: Named for Mr. Leonard Hill of Miami, Florida, who kindly donated the holotype.

Discussion: The *Murexiella macgintyi* species complex of the western Atlantic is composed of four species with disjunct ranges. These four species are separated from each other by large distances where no other shallow water *Murexiella* species are found. These four closely related and widely-separated species are probably the remnants of a once wide-ranging ancestral species which has survived, and speciated, in only four areas. The distributional pattern of this species complex is as follows; *Murexiella macgintyi* is found in southern and western Florida, the northern Bahamas, and northern Cuba; *Murexiella hilli* is only found on Aruba in the Netherlands Antilles; *Murexiella leonardhilli* is found along the northern coast of Brazil, from Pará to Salvador, Bahia; and *Murexiella taylorae* is found in the Gulf of Mexico. Between these widely separated areas, no other shallow water *Murexiella macgintyi*-type muricids are found, showing that they are, indeed, geographically isolated.

Murexiella leonardhilli differs from *M. macgintyi* in being a much larger shell with more flattened, angled whorls, in being much wider across the shoulder region, in having a longer siphonal canal with a more sharply-defined break between the siphonal canal and body whorl, and in having thicker varices with smaller, poorly-developed spines. *Murexiella leonardhilli* does not appear to vary much in color. The holotype, paratype, and several specimens that I observed in collections while in Brazil, were all a pale pinkish-orange in color.

Family **Olividae**
Subfamily **Olivinae**
Genus *Oliva* Bruguière, 1798
Subgenus *Strephona* Mörch, 1852

Oliva (Strephona) circinata tostesi new subspecies (Plate 28, Figures 3, 4)

Material Examined: HOLOTYPE - Length 41 mm, 1 m depth on sand bottom off Ilha Cabra, Ilhabela, São Paulo State, Brazil, 1986, USNM 859865; PARATYPES - 4 specimens, same locality and depth as holotype, Petuch collection; 4 specimens, same locality as holotype, Leonard Hill collection.

Description: Shell smaller than nominate subspecies, stocky, with convex, rounded sides; spire relatively low for subgenus; shell with distinct shoulder; shoulder rounded; columella with numerous plicae; color dark orange or yellowish-tan with large areas of bright blue-green, especially along shoulder; tan and blue-green base color overlaid with 2 wide bands of flammules and small triangles, one slightly posterior of mid-body, one around anterior one-third of shell; mid-body with orange-tan band with brown triangles; anterior end of columella often colored purple-brown; aperture white; inner edge of lip with line of dark purple-brown flammules; protoconch and spire enamel purple-brown.

Etymology: Named for Dr. Luiz Roberto Tostes of Rio de Janeiro, Brazil.

Discussion: *Oliva circinata tostesi* represents the southern, cold water subspecies of the more northern, tropical *O. circinata* Marrat (Plate 28, Figures 1, 2). The nominate subspecies is common along the tropical shorelines of northeastern Brazil, ranging from Rio Grande do Norte State south to Espiritu Santo State (PETUCH and SARGENT, 1986). I found *O. circinata* to be especially common in the Abrolhos Archipelago and reef complex off Bahia State, where it lived in large sand patches between reefs. *Oliva circinata tostesi*, on the other hand, is a cold water shell, inhabiting the Paulinian Province region from São Paulo State south to Santa Catarina State, where it lives on sandy-gravel bottoms in shallow protected bays. It is the southernmost species of *Oliva* known from the western Atlantic. The rocky coastlines of southern Espiritu Santo and Rio de Janeiro States may act as an ecological barrier between the two subspecies.

Oliva circinata tostesi differs from *Oliva circinata circinata* in being a much squatter, fatter, and more rotund shell with a distinct shoulder, in having a lower spire, in being smaller, and in being darker-colored, with blue-green patches, purple columella, and dark brown triangles. I had originally (PETUCH and SARGENT, 1986) referred to this shell as *O. graphica* Marrat, and used *O. circinata* as a color form of *O. graphica*. After publication of the book, and more study of Marrat's material, I now feel that the proper name for the Brazilian species is *O. circinata* with *O. graphica* being a color form of the Venezuelan *O. tisiphona* Duclos.

Family **Conidae**

Genus *Conus* Linnaeus, 1758

Conus abrolhosensis new species (Plate 28, Figures 7, 8)

Material Examined: HOLOTYPE - Length 12 mm, trawled by commercial fishermen from 20 m depth off Parcel das Paredes, Abrolhos Archipelago, Bahia State, Brazil, 1979, USNM 859872; PARATYPE - 1 specimen, same locality and depth as holotype, Kevan Sunderland collection; 1 specimen Kremer collection.

Description: Shell small for genus, biconic in shape; spire elevated; body whorl shiny, polished, sculptured with numerous raised spiral cords; spiral cords become stronger at anterior end; shoulder sharp-angled, carinated; carina coronated on spire, with coronations becoming less developed and obsolete on final half of body whorl; aperture narrow; shell color variable, ranging from orange (holotype), to white and blue-purple; holotype with scattered white patches around mid-body and anterior tip; spire whorls of holotype dark orange with evenly-spaced, oval-shaped white flammules around edge of periphery; on white and purple specimens, spire dark brown, with same pattern of evenly-spaced white flammules; protoconch large, mamillate; periostracum thin, translucent, smooth, with row of small tufts along shoulder carina and spire carina that correspond to shoulder coronations.

Etymology: Named for Abrolhos Archipelago and reef complex, Bahia State, Brazil, the type locality.

Discussion: At first glance, *Conus abrolhosensis* appears to be a member of the *C. jaspideus* Gmelin species complex from father north in the Caribbean, but it differs in having a coronated spire made up of small beads on a sharp-edged carina, and in having a distinctive, consistent, astraeiform color pattern on the spire. Based on the coronated spire and carina, and the color pattern of the spire, *C. abrolhosensis* appears to be a high-spired member of the *C. cardinalis* Hwass complex from coral reef habitats within the Caribbean region. *Conus abrolhosensis* was trawled from a coral rubble bottom, so it probably lives in a similar hard substrate environment as does its more northern relatives. The new Brazilian species is the smallest member of the *C. cardinalis* species complex.

I had originally referred this small species to *C. jaspideus* (PETUCH, 1979, p. 525), but upon closer examination, have decided that it represents a new species not related to the *C. jaspideus* group. Only two other cones are known from the Abrolhos Archipelago (the southernmost reef complex in the western Atlantic), these being *Conus iansa* Petuch (endemic to the archipelago) and *C. beddomei* Sowerby (=*C. brasiliensis* Clench). Members of the *C. jaspideus* complex have not yet been collected on the reefs of the Abrolhos Archipelago.

PLATE 27. SELDOM-SEEN GASTROPODS FROM BRAZIL

Figures 1, 2: *Latirus ogum* Petuch, 1979, holotype, 40 mm, Abrolhos Archipelago.

Figures 3, 4: *Lyria (Plicoliva) zelindae* (Petuch, 1979), holotype, 31 mm, Abrolhos Archipelago (originally placed in Olividae, but now found to belong in Volutidae).

Figure 5: *Lyria (Plicoliva) zelindae* (Petuch, 1979) 27 mm specimen, Abrolhos Archipelago.

Figures 6, 7: *Latirus vermeiji* Petuch, 1986, holotype, 25 mm, Fernando de Noronha Island.

Figures 8, 9: *Conus iansa* Petuch, 1979, holotype, 13 mm, Abrolhos Archipelago.

Figures 10, 11: *Dermomurex (Trialatella) oxum* Petuch, 1979, holotype, 13 mm, Abrolhos Archipelago.

Figures 12, 13: *Fulgurofusus ecphoroides* Harasewych, 1983, holotype, 18 mm, 264 m depth off Pará State, Brazil.

Figures 14, 15: *Conus tostesi* Petuch, 1986, holotype, 35 mm, off Cabo Frio, Rio de Janeiro State.

Figure 16: *Nodicostellaria lixa* (Petuch, 1979), holotype, 12 mm, Abrolhos Archipelago.

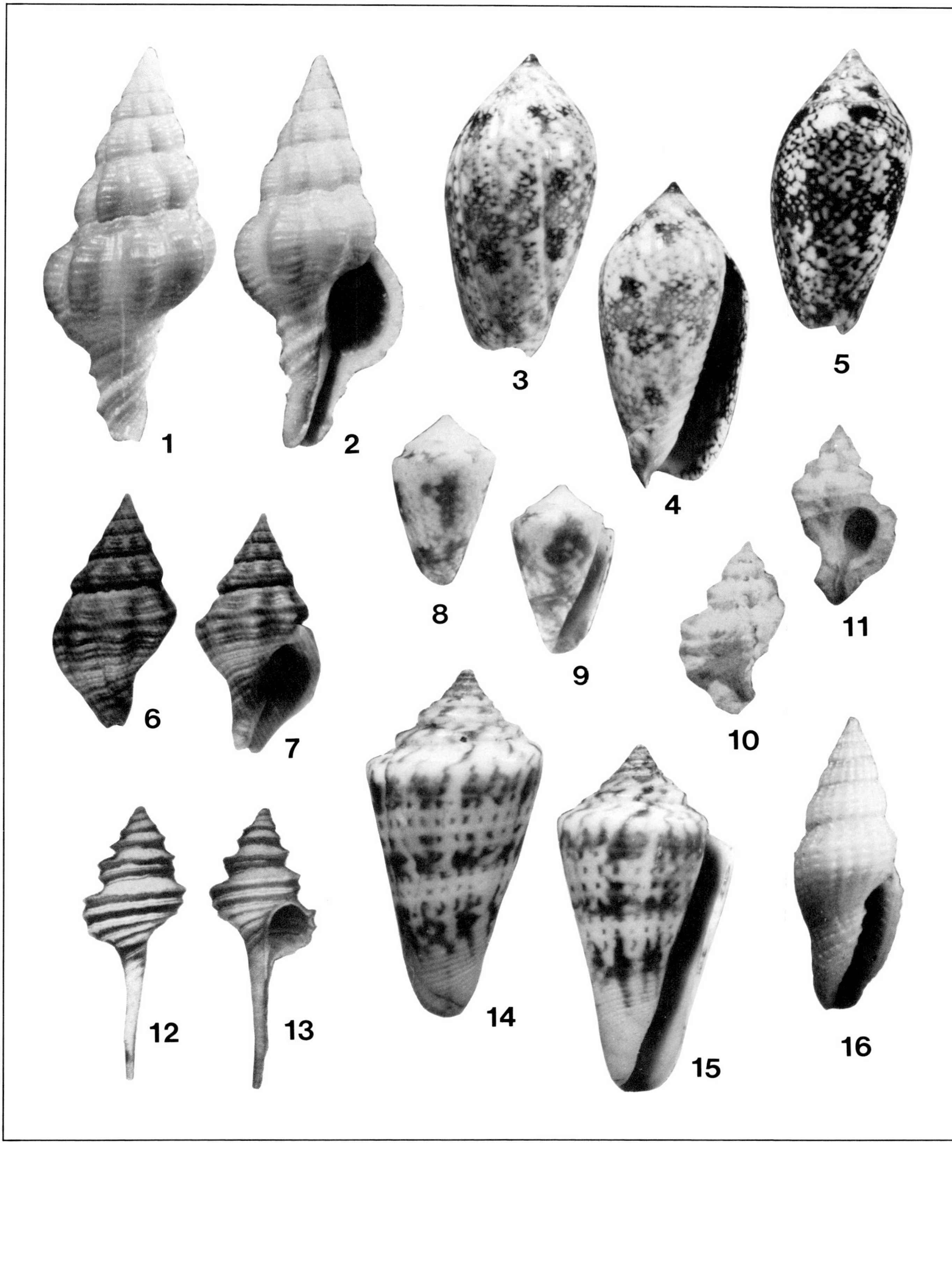
1
2
3
4
5
6
7
8
9
10
11
12
13
14
15
16

PLATE 28. NEW AND SELDOM-SEEN GASTROPODS FROM BRAZIL

Figures 1, 2: *Oliva circinata* Marrat, 1870, 42 mm specimen from Salvador, Bahia State.

Figures 3, 4: *Oliva circinata tostesi* Petuch, n.subsp., dorsal and ventral views of holotype, 41 mm.

Figures 5,6: *Murexiella leonardhilli* Petuch, n.sp., dorsal and ventral views of holotype, 31 mm.

Figures 7, 8: *Conus abrolhosensis* Petuch, n.sp., dorsal and ventral views of holotype, 12 mm.

Figures 9, 10: *Conus cf. latifasciatus* Sowerby, 1858, 44 mm specimen from off Vitoria, Espiritu Santo State.

Figures 11,12: *Conus pusillus* Lamarck, 1810, 16 mm specimen from Cabo Frio, Rio de Janeiro State.

Figures 13, 14: *Conus xanthocinctus* Petuch, 1986, holotype, 47 mm, off Cabo Frio, Rio de Janeiro State.

Figures 15, 16: *Conus carioca* Petuch, 1986, holotype, 52 mm, off Cabo Frio, Rio de Janeiro State.

Figure 17: *Conus clerii* Reeve, 1845, 46 mm specimen from off Cabo Frio, Rio de Janeiro State.

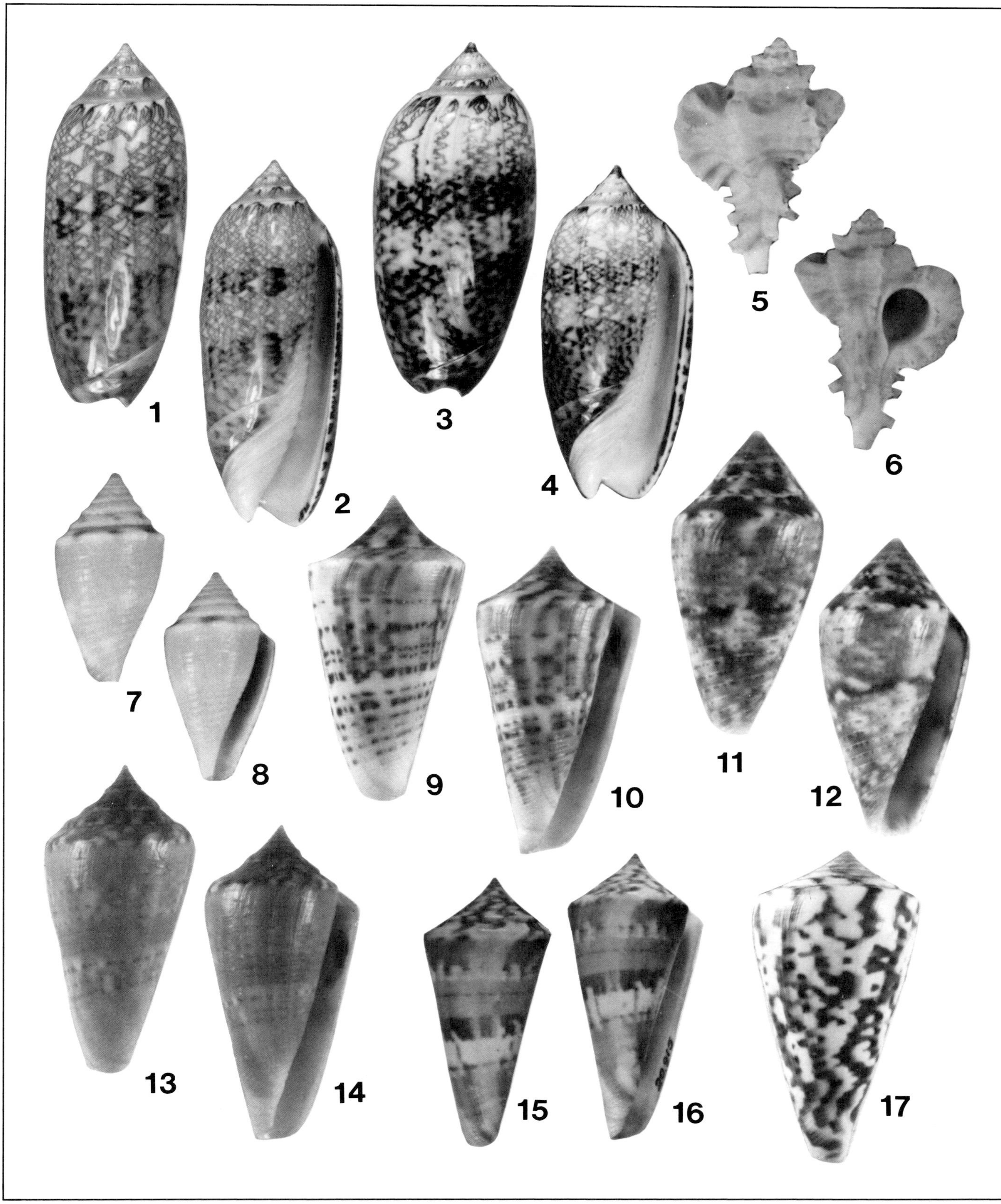
1
2
3
4
5
6
7
8
9
10
11
12
13
14
15
16
17

BIBLIOGRAPHY

Abbott, R. T., 1974. *American Seashells.* Second Edition. Van Nostrand Reinhold Company, New York. 663 pp.

Bartsch, P., 1934. *New Mollusks of the Family Turritidae* [sic]. Smithsonian Miscellaneous Collections, vol. 91, no. 2 (Johnson Fund). 29 pp., 8 plates.

Bayer, F. M., 1971. New and unusual mollusks collected by *R/V John Elliot Pillsbury* and *R/V Gerda* in the tropical western Atlantic. *Bulletin of Marine Science* 21 (1):111-236.

Beu, A. G. and Ponder W. F., 1979. A revision of the species of *Bolma* Risso, 1826 (Gastropoda: Turbinidae). *Records of the Australian Museum* 32 (1): 1-68.

Burgess, C. M., 1970. *The Living Cowries.* A. S. Barnes and Company, Inc., Cranbury, New Jersey.

Burgess, C. M., 1985. *Cowries of the World.* Gordon Verhoef Seacomber Publications, Cape Town, South Africa. 289pp.

Cate, C. N., 1973. Systematic revision of the Recent Cypraeid family Ovulidae (Mollusca: Gastropoda). *The Veliger* 15 (Supplement). 116 pp.

Cate, C. N., 1979. *A Review of the Triviidae (Mollusca: Gastropoda).* Memoir 10, San Deigo Society of Natural History. 126pp.

Dall, W. H., 1881. Preliminary report on the Mollusca. Reports on the results of dredging in the Gulf of Mexico, and in the Caribbean Sea, 1877-1879, by the United States Coast Survey steamer *Blake. Bulletins of the Museum of Comparative Zoology of Harvard* 9 (2):33-144.

Dall, W. H., 1889. Report on the Mollusca PartII. Gastropoda and Scaphopoda. Reports on the results of dredging in the Gulf of Mexico, and in the Caribbean Sea, 1879-1880, by the United States Coast Survey steamer *Blake. Bulletins of the Museum of Comparative Zoology of Harvard* 18: 1-492.

daMotta, A. J. and Harland, W., 1986. A new *Conus* (Prosobranchia: Conidae) species found in the Bahamas. *Publicaçoes Ocasionais da Sociedade Portuguesa de Malacologia* 6: 19-22.

Emerson, W. K. and W. E. Old. 1965. New records for *Cypraea surinamensis. The Nautilus* 79(1): 26-30.

Fair, R. H., 1976. *The Murex Book: An Illustrated Catalogue of Recent Muricidae.* Privately printed by Ruth H. Fair, Sturgis Printing Co., Honolulu, Hawaii.

Harasewych, M. G., 1983. A review of the Columbariinae (Gastropoda: Turbinellidae) of the western Atlantic with notes on the anatomy and systematic relationships of the subfamily. *Nemouria* No. 27, 42pp.

Harasewych,M. G. and Jensen, R. H., 1979. Review of the subgenus *Pterynotus* (Gastropoda: Muricidae) in the western Atlantic. *Nemouria* No. 22, 16pp.

Harasewych, M. G. and Petuch, E. J., 1980. *Sassia lewisi,* a new Cymatiid gastropod from the Caribbean Sea. *The Nautilus* 94(3): 121-122.

Jousseaume, F., 1869. Description d'une espece nouvelle de Cypraea. *Revue de Magasin de Zoologie,* series 2(21): 348-350, Pl. 18, figures 1-3.

Kaicher, S. D., 1982. *Card Catalogue of World-Wide Shells.* Pack No. 33, Olividae Part I.

Keen, A. Myra., 1971. *Sea Shells of Tropical West America.* Second Edition. Stanford University Press, Stanford, California. 1064pp.

Kosuge, S. and Suzuki, M., 1985. *Illustrated Catalogue of* Latiaxis *and Its Related Groups.* Special Publication No. 1, Institute of Malacology of Tokyo. 83pp.

Olsson, A. A., 1971. Mollusks from the Gulf of Panama collected by *R/V John Elliot Pillsbury,* 1967. *Bulletin of Marine Science* 21(1):35-92.

Petuch, E. J., 1974. A new *Terebra* from the coral reef areas off North Carolina. *The Veliger* 17(2):205-208.

Petuch, E. J., 1979. A new species of *Siphocypraea* (Gastropoda: Cypraeidae) from northern South America with notes on the genus in the Caribbean. *Bulletin of Marine Science* 29(2): 216-225.

Petuch, E. J., 1979. New gastropods from the Abrolhos Archipelago and Reef Complex. *Proceedings of the Biological Society of Washington* 92(3): 510-526.

Petuch, E. J., 1980. A new *Falsilyria* (Volutidae) and a new *Conus* (Conidae) from Roatan Island, Honduras (Atlantic). *The Nautilus* 94(3): 115-118.

Petuch, E. J., 1980. A new species of *Conus* from southeastern Florida (Mollusca: Gastropoda). *Proceedings of the Biological Society of Washington* 93(2): 299-302.

Petuch, E. J., 1980. A new *Falsilyria* (Volutidae) and a new *Conus* (Conidae) from Roatan Island, Honduras (Atlantic). *The Nautilus* 94(3): 115-118.

Petuch, E. J., 1980. A new species of *Conus* from southeastern Florida (Mollusca: Gastropoda). *Proceedings of the Biological Society of Washington* 93(2): 299-302.

Petuch, E. J., 1981. A relict Caenogastropod fauna from northern South America. *Malacologia* 20(2): 307-347.

Petuch, E. J., 1981. A Volutid species radiation from northern Honduras, with notes on the Honduran Caloosahatchian Secondary Relict Pocket. *Proceedings of the Biological Society of Washington* 94(4): 1110-1130.

Petuch, E. J., 1982. Geographical heterochrony: contemporaneous coexistence of Neogene and Recent molluscan faunas in the Americas. *Palaeogeography, Palaeoclimatology, and Palaeoecology* 37(1982): 277-312.

Petuch, E. J., 1982. Paraprovincialism: remnants of paleoprovincial boundaries in Recent marine molluscan provinces. *Proceedings of the Biological Society of Washington* 95(4): 774-780.

Petuch, E. J., 1986. New South American gastropods in the genera *Conus* (Conidae) and *Latirus* (Fasciolariidae). *Proceedings of the Biological Society of Washington* 99(1): 15-16.

Petuch, E. J. and Sargent, D. M., 1986. *Atlas of the Living Olive Shells of the World.* Coastal Education and Research Foundation, Charlottesville, Virginia. 253pp.

Rios, E. C., 1975. *Brazilian Marine Mollusks Iconography.* Fundaçao Universidade do Rio Grande, Centro de Ciecias do Mar, Museu Oceanografico. 331pp., 91 plates.

Rosewater, J., 1976. William Healey Dall - the legacy he left for malacology. *Bulletin of the American Malacological Union for 1975:* 4-6.

Stanley, S. M., 1986. Anatomy of a regional mass extinction: Plio-Pleistocene decimation of the western Atlantic bivalve fauna. *Palaios* 1: 17-36.

Stanley, S. M. and L. Campbell, 1981. Neogene mass extinction of western Atlantic molluscs. *Nature* 293 (8 October): 457-459.

Sowerby, J. B. II, 1880. *Thesaurus Conchyliorum or Monographs on Genera of Shells,* Vol. 4. London.

Taylor, J. and J. G. Walls, 1975. *Cowries.* T. F. H. Publications Inc., Ltd., Neptune City, New Jersey. 288pp.

Vermeij, G. J. and E. J. Petuch, 1986. Differential extinction in tropical American molluscs: endemism, architecture, and the Panama Land Bridge. *Malacologia* 27(1): 29-41.

Vokes, E. H., 1974. Notes on the fauna of the Chipola Formation XV - On the occurrence of *Ranella poppelacki* Hörnes, a gastropod of uncertain affinities. *Tulane Studies In Geology and Paleontology* 11(2): 96-98.

Wagner, R. J. L., and R. T. Abbott, 1978. *Standard Catalog of Shells.* Third Edition with supplements. American Malacologists, Inc., Greenville, Delaware.

Walls, J. G., 1979. *Cone Shells, A Synopsis of the Living Conidae.* T. F. H. Publications, Inc., Ltd., Neptune City, New Jersey. 1011pp.

SYSTEMATIC INDEX

Numbers in bold type indicate plates on which specimens are illustrated; numbers in italic type indicate the figure numbers on the given plate showing the specimen.

Numbers in bold type indicate plates on which specimens are illustrated; numbers in italic type indicate the figure numbers on the given plate showing the specimen.

Numbers in bold type indicate plates on which specimens are illustrated; numbers in italic type indicate the figure numbers on the given plate showing the specimen.

Numbers in bold type indicate plates on which specimens are illustrated; numbers in italic type indicate the figure numbers on the given plate showing the specimen.

Numbers in bold type indicate plates on which specimens are illustrated; numbers in italic type indicate the figure numbers on the given plate showing the specimen.

ADDENDUM

While this book was in press, two new Florida gastropods were brought to my attention. These new species are important components of the deeper water ecosystems of the eastern Gulf of Mexico and Florida Keys and needed to be described. The descriptions of a new muricid and a new fasciolariid are here appended to the book.

Family **Muricidae**
Genus *Murex* Linnaeus, 1758

Murex lindajoyceae new species (Plate 29, Figures 1, 2, 3)

Material Examined: HOLOTYPE - Length 29 mm, trawled by deep water shrimpers (Royal Red Shrimp) from 150 m depth due west of Cedar Key, Florida, 1984, USNM; PARATYPE - Length 29 mm, same locality and depth as holotype, USNM; 4 specimens, lengths 25 mm to 29 mm, same locality and depth as holotype, Petuch collection.

Description: Shell very small, thin, and delicate for genus; body whorl rounded, compressed; 3 thin varices per whorl; varices extend onto siphonal canal; 3 straight, long, needle-like spines per varix on body whorl; 1 or 2 spines on siphonal extension of varix; intervarical areas with 3, or occasionally 4, rounded, beaded, axial ribs; siphonal canal straight, very elongated and delicate, more than twice the length of the body whorl; aperture large in proportion to shell size, oval in shape; shell color pale salmon-pink with 2 bright red bands, one around shoulder and one at junction of body whorl and siphonal canal; siphonal canal of some specimens dark reddish brown; protoconch and early whorls dark salmon-orange.

Etymology: Named for my wife, Linda Joyce Petuch.

Discussion: *Murex lindajoyceae* is a faunal component of the Tampan Caloosahatchian Relict Pocket of the northwestern Florida continental shelf. This delicate and colorful little species has frequently been misidentified by collectors and shell dealers as *Murex anniae* M. Smith, 1940. That species, however, is a fossil from the Caloosahatchee Formation of southern Florida, and is a much larger and more robust shell with only two intervarical ribs. *Murex lindajoyceae* also has been confused with the similar-appearing *M. tryoni* Hidalgo, 1880, but differs in having a proportionately smaller body whorl, in having fewer and more coarsely-sculptured intervarical axial ribs, and in being pink with two red bands instead of pure white. When trawling in the 100-200 m depth range along northwestern Florida, *Murex lindajoyceae* is frequently collected together with *Chicoreus rachelcarsonae* Petuch and *Fasciolaria bullisi* Lyons.

Family **Fasciolariidae**
Subfamily **Peristerniinae**
Genus *Leucozonia* Gray, 1847

Leucozonia jacarusoi new species (Plate 29, Figures 4, 5)

Material Examined: HOLOTYPE - Length 24 mm, brought up in deep water lobster trap from 70 m depth off Key West, Florida, 1986, USNM; PARATYPES - Length 16 mm, same locality and depth as holotype, USNM; 10 specimens, same locality as holotype, Kevan Sunderland collection.

Description: Shell thin for genus, lightweight, fusiform in shape; spire elevated, exserted; 8-10 sharp-angled knobs on shoulder of each whorl; shoulder knobs ornamented with 3 large spiral cords; large cord around mid-body which terminates in small spine on outer lip; siphonal canal long, well-developed, ornamented with two large spiral cords and numerous fine spiral threads; columella with 3 or 4 small plications; aperture with several beaded cords; shell color uniformly pure white; interior aperture pure white; occasional specimens uniformly pale yellow.

Etymology: Named for Mr. James Jacarusco of Stuart, Florida, who collected the type lot from lobster traps.

Discussion: At first glance, *Leucozonia jacarusoi* somewhat resembles an albinistic *L. nassa* (Gmelin, 1791) (Plate 29, Figure 6) but differs in a number of consistent ways: The new species is always white or pale yellow while *L. nassa* is always dark brown or black with a white band; the shoulder angle of *L. jacarusoi* is much sharper than that of *L. nassa,* with the shoulder knobs being more pronounced and acute; the siphonal canal of the new species is much longer and better developed than that of *L. nassa* and is encircled by two large cords; the mid-body cord of *L. jacarusoi* is smaller and much more indented than that that of *L. nassa* which, together with the shoulder knobs, has a bicarinated appearance in profile; the new species is smaller, more elongated, and much more lightweight than *L. nassa;* the outline of *L. jacarusoi,* with its more indented mid-body and sharper-angled shoulder, more closely resembles a *Latirus* than a typical *Leucozonia.*

Ecologically, the two species are very different. *Leucozonia nassa* prefers rubble areas on rocky shorelines and reefs, where it is primarily an intertidal or shallow subtidal species, and is rarely found below 6 meters depth. *Leucozonia jacarusoi* is found in depths of around 70 meters (possibly deeper), where it most probably lives on the coralline algal nodule bottoms north of the Florida Keys and along the outer edge of the western Florida continental shelf. The deep water *Murexiella kalafuti* Petuch is frequently collected in lobster traps with the new *Leucozonia.* Off the Florida Keys, the lobster traps are often placed on these coralline algal bottoms in deep water. Both *Murexiella kalafuti* and *Leucozonia jacarusoi* appear to be attracted to the traps by barnacles, their prey item, which often heavily encrust the traps.

Zoogeographically, the two *Leucozonia* species also differ. *Leucozonia nassa* ranges from Florida and the Gulf of Mexico southward to central Brazil, where it is almost always an abundant intertidal species. No deep water (50 meters or more) morphs of *L. nassa* are known from anywhere along this range. *Leucozonia jacarusoi,* on the other hand, has only been collected from deep water (70 meters) off the Florida Keys, and the species is probably restricted to the coralline algal bottoms of southern and southwestern Florida.

PLATE 29

Figures 1, 2: *Murex lindajoyceae* new species, dorsal and ventral views of holotype, 29 mm.

Figure 3: *Murex lindajoyceae* new species, dorsal view of paratype, length 29 mm.

Figures 4, 5: *Leucozonia jacarusoi* new species, dorsal and ventral views of holotype, length 24 mm.

Figure 6: *Leucozonia nassa* (Gmelin, 1791), dorsal view of 29 mm specimen.

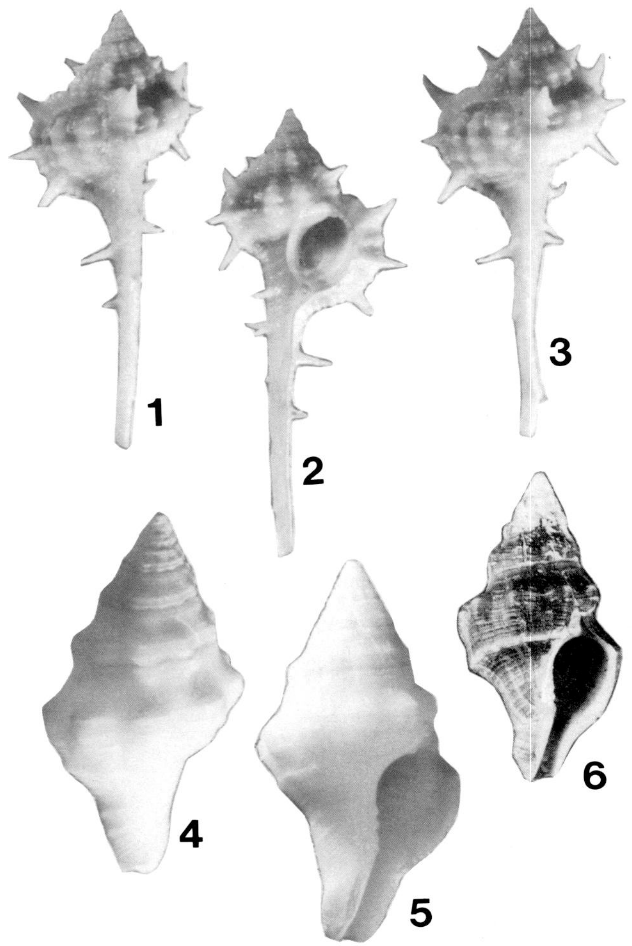
1
2
3
4
5
6